Plasma Science - Recent Advances, New Perspectives and Applications

Edited by Sukhmander Singh

Published in London, United Kingdom

Plasma Science - Recent Advances, New Perspectives and Applications
http://dx.doi.org/10.5772/intechopen.104018
Edited by Sukhmander Singh

Contributors
Hirak Jyoti Dehingia, Mohamed Himdi, Fatemeh Sadeghikia, Ali Karami Horestani, Jyoti, Hector Pérez-de-Tejada, R. Lundin, Mukesh Kumar Alaria, Sukhmander Singh, Bhavna Vidhani, Sonia Yogi, Ashish Tyagi, Sanjeev Kumar, Shravan Kumar Meena

First published in London, United Kingdom, 2023 by IntechOpen
IntechOpen is the global imprint of INTECHOPEN LIMITED, registered in England and Wales, registration number: 11086078, 5 Princes Gate Court, London, SW7 2QJ, United Kingdom

British Library Cataloguing-in-Publication Data
A catalogue record for this book is available from the British Library

Additional hard and PDF copies can be obtained from orders@intechopen.com

Plasma Science - Recent Advances, New Perspectives and Applications
Edited by Sukhmander Singh
p. cm.
Print ISBN 978-1-83768-023-8
Online ISBN 978-1-83768-024-5
eBook (PDF) ISBN 978-1-83768-025-2

Meet the editor

Dr. Sukhmander Singh has been an assistant professor in the Department of Physics, Central University of Rajasthan, Ajmer, India, since 2015. From 2012 to 2015, he worked as an assistant professor at Motilal Nehru College, University of Delhi, India. He obtained his MSc in Physics from Jawaharlal Nehru University New Delhi and Ph.D. in Plasma Physics from the Indian Institute of Technology Delhi (IIT Delhi), New Delhi, India. He has published forty research articles and many conference proceedings in journals of international repute. He has contributed ten book chapters and edited three books. He is an active reviewer of many international journals. His area of interest is the theory and simulation of plasma waves and instabilities.

Contents

Preface **IX**

Section 1
Plasma Waves and Instabilities **1**

Chapter 1 **3**
Introductory Chapter: Recent Advances in Plasmas Physics
by Sukhmander Singh

Chapter 2 **7**
Plasma Waves and Rayleigh–Taylor Instability: Theory and Application
by Sukhmander Singh, Bhavna Vidhani, Sonia Yogi, Ashish Tyagi, Sanjeev Kumar and Shravan Kumar Meena

Section 2
Nonlinear Phenomena in Plasma **29**

Chapter 3 **31**
Various Aspects of Dust-Acoustic Solitary Waves (DAWs) in Inhomogeneous Plasmas
by Hirak Jyoti Dehingia

Chapter 4 **45**
A Summary of the Introduction and Importance of Quantum Plasmas
by Jyoti

Section 3
Particle Acceleration in the Venus Wake **57**

Chapter 5 **59**
Particle Acceleration in the Venus Plasma Wake
by Hector Pérez-de-Tejada and R. Lundin

Section 4
Plasma Microwave Interactions **71**

Chapter 6 **73**
Design and Analysis of High Power RF Window
by Mukesh Kumar Alaria

Chapter 7 **89**
Reconfigurable Antennas Based on Plasma Reflectors and Cylindrical Slotted Waveguide
by Fatemeh Sadeghikia, Ali Karami Horestani and Mohamed Himdi

Preface

Plasma refers to a state of matter or ionized gas consisting of a collection of charged particles, including electrons and positive ions, along with neutral particles. These charged particles are coupled via electromagnetic interaction that exhibits collective behavior. This ionization can occur through various processes such as external heating, collisions, or exposure to strong electromagnetic fields via laser-plasma interactions. The naturally occurring plasmas include stars, lightning, and the Earth's ionosphere and can also be generated artificially in laboratories and industries for materials synthesis and surface modification. Plasma is a highly conductive substance, which makes it useful for many applications such as in the creation of plasma televisions, fluorescent lights, and fusion reactors.

This book discusses linear and nonlinear aspects of plasma physics. The chapters present research in the field and discuss such topics as waves and instabilities in plasma and their dispersion, technological aspects of plasma devices, derivation, and more. This book is a useful resource for postgraduate students and researchers in plasma physics.

Chapter 1 presents a summary of the book, while Chapter 2 deals with stable and unstable plasma oscillations with the help of two-fluid plasma hydrodynamics under the linear wave analysis for small-amplitude waves (nonlinear terms are ignored). It also discusses the applications of Rayleigh-Taylor instability in different fields. The dispersion relation is solved numerically to discuss the growing waves in plasma.

Chapter 3 reviews nonlinear wave structures such as solitons, supersolitons, rouge waves, shock waves, and others. The nonlinearities cannot be overlooked for the high-amplitude waves, as mentioned in Chapter 2. Chapter 3 also discusses the various properties of dust lattice solitary waves, dust acoustic solitary waves, shock waves, and dust ion-acoustic solitary waves in an inhomogeneous plasma.

Chapter 4 investigates the growth rate of plasma instability in quantum plasma under the normal mode analysis for small-amplitude waves. The thermal de Broglie wavelength of ions and electrons is comparable to the inter-particle separation in dense plasmas that are kept at low temperatures. Under these conditions, the quantum effect becomes more significant than the classical one.

Chapter 5 explains particle acceleration in the Venus plasma wake.

Chapter 6 covers the selection of dielectric materials and power handling capability of radiofrequency (RF) windows. The gyrotron is used in plasma fusion machines as a high-power millimeter-wave source for electron cyclotron resonance heating. The design analysis of RF windows with window disk thickness and diameter are optimized considering the minimum return loss and the minimum insertion loss by using CST MICROWAVE STUDIO software.

Chapter 7 focuses on the application of plasma structures for the realization of reconfigurable antennas.

I extend my heartfelt gratitude to our contributing authors Dr. Jyoti, Dr. Dehingia, Dr. Mukesh, Dr. Tejada, Dr. Lundin, Dr. Sadeghikia, Dr. Horestani, and Dr. Mohamed for their high-quality chapters. Their dedication and insights have enriched the content of the book and made it possible to deliver valuable information to our readers. I am also thankful to Author Service Manager Mr. Dominik Samardzija and the staff at IntechOpen for their effort and support. Special thanks to my parents and sisters as well as my wife Rajvinder Kaur for her cooperation in my life and for sharing the responsibility of family affairs so that I could spend my time on this book. A special mention goes to my daughter Jashandeep whose presence brings me immense joy and inspiration.

We would be pleased to receive comments, suggestions, and/or criticisms from the readers for further improvements.

Sukhmander Singh
Plasma Waves and Electric Propulsion Laboratory,
Department of Physics,
Central University of Rajasthan,
Ajmer, Kishangarh, India

Section 1

Plasma Waves and Instabilities

Chapter 1

Introductory Chapter: Recent Advances in Plasmas Physics

Sukhmander Singh

1. Introduction

Plasma, the fourth state of matter, is a highly ionized form of gas where the density of ions and free electrons is high enough to exhibit collective behavior. In a plasma, the ions and free electrons are no longer bound to individual atoms or molecules but instead form a collective "soup" of charged particles. In a plasma, the behavior of charged particles is governed by both electromagnetic forces and interparticle collisions, including Van der Waals forces. However, in many cases, the effects of electromagnetic fields are dominant over the effects of interparticle collisions [1]. These electromagnetic interactions can result in collective behavior, such as the formation of waves and oscillations in the plasma. In a plasma, waves can be attenuated due to a number of mechanisms, such as collisional damping, Landau damping, and cyclotron damping. The plasma can be created using a variety of methods, including microwave discharges, radio frequency (RF) discharges, and arc discharges. The properties of the plasma can be controlled by adjusting the power, frequency, and duration of the discharge [2].

2. Concept of plasma fluid instabilities

Plasma fluid instabilities are a common occurrence in plasma physics, where they play a critical role in the behavior and dynamics of plasmas. These instabilities can be classified into two broad categories: hydrodynamic instabilities, which are driven by fluid dynamics, and kinetic instabilities, which are driven by the interactions between individual particles. Some examples of plasma fluid instabilities include the Rayleigh-Taylor instability, the Kelvin-Helmholtz instability, the tearing mode instability, and the drift wave instability. These instabilities can have significant impacts on plasma behavior, including transport, confinement, and energy transfer. For instance, the Rayleigh-Taylor instability can lead to plasma mixing and turbulence, while the Kelvin-Helmholtz instability can cause plasma transport across magnetic field lines. Plasma has numerous industrial applications due to its unique properties such as high energy, high temperature, and chemical reactivity [3–14].

3. Applications of plasma in various fields

Plasma is used in the manufacturing of semiconductors, which are used in electronic devices. The plasma etching process is used to pattern silicon wafers for the

IntechOpen

production of integrated circuits. Plasma technology is used in aerospace and defense applications, such as plasma thrusters for space propulsion, plasma-assisted combustion for jet engines, and plasma-based weapons.

Plasma antennas work by ionizing gas to create a plasma, which can conduct electromagnetic waves. A plasma antenna uses plasma as the conducting medium instead of metal. Plasma antennas have several advantages over traditional metal antennas, including their ability to be reconfigured in real time, their low profile, and their ability to operate over a wide range of frequencies. Today, plasma antennas are used in a military and civilian communication systems, radar, and satellite communication.

The study of quantum plasma is an exciting and rapidly evolving area of research that has the potential to revolutionize our understanding of the behavior of matter at the quantum level. The quantum effects become important because the length and timescales involved are on the order of the de Broglie wavelength. This leads to the emergence of collective modes, such as plasmons and phonons, which are not observed in classical plasmas. The modes are quantum plasma wakefield acceleration and Bose-Einstein condensation (leading to novel behavior such as the superfluidity of the plasma). The quantum plasma is relevant for understanding the behavior of plasmas in extreme environments such as the interiors of stars and the early universe.

In plasmas, solitary waves can form due to the nonlinear behavior of the plasma, where the response of the plasma to an external perturbation depends on the amplitude of the perturbation, and the response of the plasma to an external perturbation depends on the amplitude of the perturbation. The solitons can be described mathematically with the Korteweg-deVries (KdV) equation. Solitary waves cab be used to transport energy and momentum in plasmas over a long distance. The references discuss the ion acoustic soliton experiments in a plasma [15–17]. A Gyrotron is a high-power, high-frequency microwave device that uses the interaction between the electrons and the cyclotron resonance frequency (CRF) to generate RF energy. Solitary waves can be used in fusion research, industrial heating, and advanced communication systems [18].

Author details

Sukhmander Singh
Plasma Waves and Electric Propulsion Laboratory, Department of Physics, Central University of Rajasthan, Ajmer, Kishangarh, India

*Address all correspondence to: sukhmandersingh@curaj.ac.in

References

[1] Bittencourt JA. Fundamentals of Plasma Physics. Springer-Verlag New York, Inc., Springer Science & Business Media; 2004

[2] Lieberman MA, Lichtenberg AJ. Principles of plasma discharges and materials processing. MRS Bulletin. 1994;**30**(12):899-901

[3] Verheest F. Waves and instabilities in dusty space plasmas. Space Science Reviews. 1996;**77**(3-4):267-302

[4] Taroyan Y, Ruderman MS. MHD waves and instabilities in space plasma flows. Space Science Reviews. 2011;**158**(2-4):505

[5] Chandrasekhar S. Hydrodynamic and Hydromagnetic Stability, Dover, 1981, First Published by. England: Oxford University Press; 1961

[6] Roberts MS. Fluid instabilities and transition to turbulence. In: Ghaedi K et al, editors. Computational Overview of Fluid Structure Interaction. London: IntechOpen; 2020. DOI: 10.5772/intechopen.92219

[7] Drazin PG. Introduction to Hydrodynamic Stability (Cambridge Texts in Applied Mathematics). United Kingdom: Cambridge University Press; 2002. DOI: 10.1017/CBO9780511809064

[8] Allah MO. An overview of linear and nonlinear Rayleigh-Taylor instability. General Mathematics Notes. 2014;**20**(1):67-76

[9] Francois C. Hydrodynamic Instabilities. United Kingdom: Cambridge University Press; 2011. DOI: 10.1017/CBO9780511975172

[10] Singh S. In: Ghaedi K Et al., Editors. Dynamics of Rayleigh-Taylor Instability in Plasma Fluids in Computational Overview of Fluid Structure Interaction. London: IntechOpen; 2020. DOI: 10.5772/intechopen.92025

[11] Singh S. In: Shahzad A, editor. Waves and Instabilities in E × B Dusty Plasma in Thermophysical Properties of Complex Materials. London: IntechOpen; 2019. DOI: 10.5772/intechopen.90397

[12] Singh S. In: Ghaedi K et al., editors. Evolutions of Growing Waves in Complex Plasma Medium in Computational Overview of Fluid Structure Interaction. London: IntechOpen; 2020. DOI: 10.5772/intechopen.93232

[13] Singh S, Tyagi A, Vidhani B. In: Wang L, editor. Physics of Absorption and Generation of Electromagnetic Radiationin Electromagnetic Wave Propagation for Industry and Biomedical Applications. London: IntechOpen; 2021. DOI: 10.5772/intechopen.99037

[14] Singh S, Vidhani B, Tyagi A. In: Shahzad A, editor. Numerical Investigations of Electromagnetic Oscillations and Turbulences in Hall Thrusters Using Two Fluid Approach in Plasma Science and Technology. London: IntechOpen; 2021. DOI: 10.5772/intechopen.99883

[15] Lonngren KE. Ion acoustic soliton experiments in a plasma. Optical and Quantum Electronics. 1998;**30**:615-630. DOI: 10.1023/A:1006910004292

[16] Lonngren KE. Soliton experiments in plasmas. Plasma Physics. 1983;**25**:943. DOI: 10.1088/0032-1028/25/9/001

[17] Harvey P, Durniak C, Samsonov D, Morfill G. Soliton interaction in a

complex plasma. Physical Review E. 2010;**81**(5):057401

[18] Speth E. Development of powerful RF plasma sources for present and future NBI systems. Plasma Science and Technology. 2004;**6**(1):2135

Chapter 2

Plasma Waves and Rayleigh–Taylor Instability: Theory and Application

Sukhmander Singh, Bhavna Vidhani, Sonia Yogi, Ashish Tyagi, Sanjeev Kumar and Shravan Kumar Meena

Abstract

The presence of plasma density gradient is one of the main sources of Rayleigh–Taylor instability (RTI). The Rayleigh–Taylor instability has application in meteorology to explain cloud formations and in astrophysics to explain finger formation. It has wide applications in the inertial confinement fusion to determine the yield of the reaction. The aim of the chapter is to discuss the current status of the research related to RTI. The current research related to RTI has been reviewed, and general dispersion relation has been derived under the thermal motion of electron. The perturbed densities of ions and electrons are determined using two fluid approach under the small amplitude of oscillations. The dispersion equation is derived with the help of Poisson's equation and solved numerically to investigate the effect of various parameters on the growth rate and real frequency. It has been shown that the real frequency increases with plasma density gradient, electron temperature and the wavenumber, but magnetic field has opposite effect on it. On the other hand, the growth rate of instability increases with magnetic field and density gradient, but it decreases with electron temperature and wave number.

Keywords: Rayleigh Taylor instability, plasma, density gradient, real frequency, growth rate

1. Introduction

In the fluid mechanics, there are mainly two types of flows, i.e., laminar and non-laminar. Streamline flow is the example of laminar flow, and turbulent flow is non-laminar. A laminar flow converted to non-laminar due to some gradient or due to some perturbation in pressure and temperature. During the process of converting a flow from laminar to non-laminar, it produces some patterns in between the two states which are unstable and caused due to instability in the fluid [1]. There are many instabilities occurred due to many reasons, for example, Rayleigh–Taylor instability, Kelvin–Helmholtz instability, electrostatic instability, drift instability, electromagnetic instability, etc. Chandrasekhar was the first to introduce viscosity while analyzing the RTI problem at spherical accelerated interface [1]. In general instability means redistribution of energy, and this redistribution can occur due to some perturbation (external small disturbance). In some cases, we need to compress/reduce the instability to make better use of the appliance or the system to improve the performance. As in case of Rayleigh–Taylor instability, a heavy fluid is loaded on top of the lighter fluid in presence of gravitation, which always acts in downward direction as shown in **Figure 1**. Rayleigh–Taylor instability is a type of buoyancy-driven instability [1–3]. One example that can be observed at home is, when cold milk is added to hot tea/water (lighter-less density), then there are spikes created in the tea/water due to milk, and this is basic example of Rayleigh–Taylor instability. Some more examples of Rayleigh–Taylor instability which occurs in nature and industrial process are the instability in cold and hot water of oceans and rivers, in ionosphere due to solar radiation, in nuclear explosion and gas bubbles formation under the water. Most of the time RTI was solved using linear equations due to small perturbation but nonlinear method also exists for the situation, when perturbed amplitude is more and nonlinear terms of the equation cannot be neglected due to their reasonable magnitude [1–5].

Another type of instability is the Kelvin–Helmholtz instability is one another macro-instability, which is due to shearing effect in single fluid due to different velocities or having different velocities at the interface of two fluids relative to each other [5]. It can be observed, when clouds make spiral shape patterns and roll around each other, then there exists Kelvin–Helmholtz Instability. It is caused due to flow of bulk plasma and have considerably long wavelength than gyro-radii of any kind [3]. Some examples include the red spot on Jupiter and the Sun's atmosphere [2]. Electrostatic instabilities K-H and R-T are caused by velocity space inhomogeneity [3]. As an electromagnetic wave can propagate in plasma, electrostatic wave cannot even leave and penetrate the plasma. The electromagnetic instability is also a micro-instability, and it is caused due to velocity-space inhomogeneity or deformation of phase-space distribution function. Both electrostatic and electromagnetic plasma are developed in an unmagnetized plasma with different specific conditions for both.

The another universal and linear instability known as drift instability is caused by destabilizing of drift waves (which is driven by pressure gradient) due to difference in motion of the two components of plasma, i.e., electrons and ions in a magnetized plasma. Generally, plasma are inhomogeneous and density gradient in plasma support drift wave, which trigger drift instability in microscopic scale. Maximum time this type of instability exists in astrophysical plasma [2, 3].

Talking about the Rayleigh–Taylor Instability, there are a lot of applications for Rayleigh–Taylor instability from space to laboratory, from plasma to any fluid. But there are certain applications that are more beneficial where Rayleigh–Taylor instability (RTI) is important to study to make good material, machine or to reduce the energy loss due to instability in some part of system [6–9]. RTI is used in tokamak to

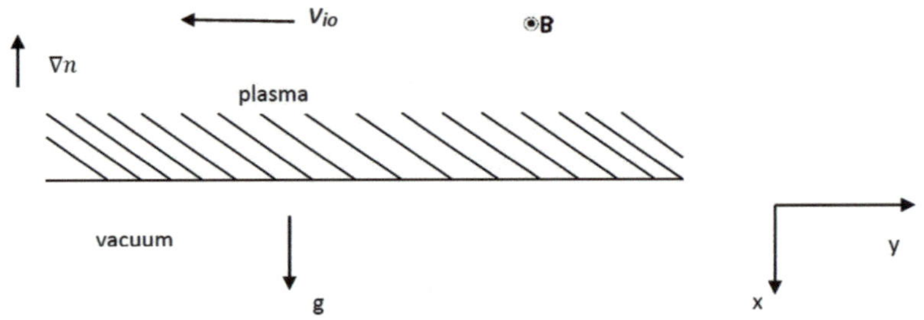

Figure 1.
Plasma species under a gravitational field.

confine the plasma and gases by giving the conditions using comparing time of instability and fusion. More different types of RTI were studied in the past with simulations in 2D and 3D and observed a RTI in case of magnetohydrodynamics in the supernova remnants. Recently, Rayleigh–Taylor instability is observed in the outer surface of the sun where plasma bubble with low density carries high density solar prominence. Rayleigh–Taylor instability occurs in interstellar gases containing magnetic field and a density gradient. In sonoluminescense phenomenon, the RTI limits driving pressure. Rayleigh–Taylor instability has a critical role in case of climate which leads to drastic drop in temperature [4]. RTI also exists in salt domes, which occurred over millions of years ago under Earth's crust under density gradient caused by acceleration due to gravity in upward direction.

2. Review of the current status

The heat and mass transfer rates of different viscosity and density of liquids can be controlled through instabilities. These instabilities lead to unstable multilayer flows, which results in nonuniform film thickness of coating processes. Hydrodynamic stability is well-known problem of fluid dynamics, which tells that when and why laminar flow break down. Zanella et al. presented the Cahn–Hilliard/Navier–Stokes model at low Atwood number to simulate the development of RTI and low-intermediate Reynolds number in immiscible fluids. The diffuse-interface model is presented in 2 D and 3D simulations capture phenomenon [10]. Balestra et al. inquire the RTI inside a cylinder, coated with a thin liquid film and axis orthogonal to gravity to investigate the effects of geometry on the instability [11]. Zeng et al. analyzed the rotational part of the disturbance flow field at the cylindrical interface of RTI and found the most unstable mode for small radii interfaces [12]. Tamim and Bostwick performed an elastodynamic stability analysis in a cylindrical container for the visco-elastic layer and tally the dispersion relationship, which shows dependence on the elastogravity number, elastocapillary number, compressibility number and solid Deborah number. They calculated the associated growth rate and wavelength by stability diagram in solid RTI [13].

Gebhard et al. described the incompressible inhomogeneous Euler equations for two different fluids under the influence of gravity and derived a general criterion for the turbulent mixing of two different fluids. It is found that for ultra-high range Atwood number the mixing grows quadratically in time [14]. Zhu et al. presented a method to run an azimuthal electron current that utilizes a rotating magnetic field to decrease the charge separation due to RTI. The azimuthal electric field of the fluctuation approaches to zero [15]. The mathematical homotopy perturbation method is employed to solve complex nonlinear differential equations. Two superposed rotating infinite hydromagnetic Darcian flows under the influence of a tangential uniform magnetic field through porous media are solved [16]. Mondal and Korenaga presented a propagator matrix formulation of stokes flow and instability problem turned into small eigenvalue equation. They use Monte Carlo sensitivity analysis and derived formula to calculate the growth rate of RTI for the stability of early crust [17]. Rahman and San investigated Euler turbulence using a central seven-point stencil reconstruction scheme for the performance of a relaxation filtering approach. For single-mode and multi-mode, a high-resolution numerical experiments are performed for inviscid RTI problems in 2D canonical settings [18].

Song and Srinivasan investigate the RTI under the effect of resistivity, magnetic field, viscosity and thermal conduction [19]. Casner et al. reviewed different RTI experiments performed at classical embedded interface or at the ablation front. A bubble-merger, bubble- competition regime in 2D was evidenced for the first time in indirect-drive on the National Ignition Facility at the ablation front all due to an unprecedented x-ray drive [20]. Jha et al. studied volume of fluid method to see the stabilizing effects of RTI simulations. The effect of numerical stability of blending parameter has been investigated on the flow topology. The numerical stability comparison of blending scheme shows better result than the pure higher-order schemes [21]. Yilmaz uses an advanced and fully featured LES algorithm to simulated RTI at high Atwood numbers [22]. Ahuja and Girotra investigated the RTI of nanofluids having different densities with superimposed horizontal layers [23].

Zhao et al. given an analytical model at arbitrary Atwood number in a cylindrical geometry for the nonlinear evolution of 2D single-mode RTI [24]. Horne and Lawrie studied the turbulent mixing induced by gravitational force between two miscible fluids in a tightly confined domain [25]. Scase and Hill argued about the effect of rotation on the classical gravity-driven RTI, which resulted between the destabilization effect of gravity and stabilization effect of the rotation. A fourth-order Orr–Sommerfeld equation was derived via linear stability analysis [26]. Chen et al. studied the coexisting system and 2D Richtmyer–Meshkov instability system. The multiple-relaxation time discrete Boltzmann model was employed for Richtmyer–Meshkov instability system, and the correlation between non-uniformity of temperature and globally averaged non-organized energy flux is found one [27]. Elyanov et al. studied the condition for the appearance of various types of instabilities in the lean hydrogen–air mixture containing inhomogeneities at the flame front [28].

Alqatarl et al. created a clean initial stationary interface between two miscible fluids by preparing a density inversion inside the thin gas separating flat plates. As a result of these conditions, the instability is suppressed below a critical plate spacing [29]. Gopalakrishnan et al. combined the mechanism of double-diffused instability and RTI for porous media flows [30]. Scase and Sengupta carried out numerical simulations of the inviscid flow, and they also examine the special case of single-layer rotating viscous column [31]. Sauppe et al. found that the initial sinusoidal perturbation grows slowly at the time of implosion and develops the classical bubble in the experiment of RTI growth (laser-driven cylindrical implosion) [32]. Shimony et al. find a relation for growth parameter of the self-similar evolution and the density of bubbles. They examined the effect of density ratio on RTI [33]. Malamud et al. studied the evolution of RTI in the design of a laser-driven experiment and examined the re-acceleration phenomenon [34]. Sauppe et al. designed a cylindrical implosion experiment using the radiation-hydrodynamics code xRAGE for the OMEGA and NIF laser facilities [35]. Wu et al. concentrated on nonlinear dynamics and analyzed the 3D single-mode RTI at a higher Atwood number [36]. Liang et al. performed the numerical simulations using an improved phase field lattice Boltzmann method at low Atwood number for multimode immiscible RTI. The larger wavelength instability undergoes faster growth after analyzed the effect of the initial conditions in terms of the perturbed amplitude and wavelength [37]. Wolf studied the miscible liquids for dynamic stabilization of RTI and the related frozen waves [38]. Cohen and Oron studied the dynamics of a thin isothermal and non-isothermal liquid film under the influence of harmonic tangential force [39]. Nimmagadda et al. studied the dynamics of a gradually mixed fluid with pre-defined bubbles rising due to RTI [40]. Naveh et al. theoretically studied the late-time evolution inside a finite-sized spatial domain

[41]. Panda et al. investigated the seasonal and Spatial-temporal effects of ionosphere irregularities on GNSS using various instances of ionospheric disturbances [42]. Hu et al. simulated the evolution in 2D single-mode RTI of the bubble vorticity and its velocity at different viscosities with different Reynold number and at low Atwood number [43].

Luo et al. numerically studied the effect of compressibility on RTI with different isothermal Atwood numbers and Mach number in the late-time evolution of 2-D single-mode [44]. Wieland et al. used wavelet-based adaptive mesh refinement to study the vorticity dynamics and the impact of isothermal stratification strength on it for compressible single-mode RTI [45]. Puillet et al. used rotating cylinder coated by liquid elastomer melt from outside, to produce an array of droplets on a thin layer of the melt to study the elastic amplification of RTI [46]. Baksht et al. considered the energy balance in the radiating metallic gas-puff Z pinch to study the energy balance in imploding Mg-shell-on-Bi-jet pinches and Mg-shell-on-Mg-jet. The power-law density distribution is observed in this type of Z pinch for promoting suppression of RTI that occurs during compression in the pinch plasma [47]. Mishra et al. analyze the observations of an eruption in the intermediate corona consisting of magnetic Rayleigh–Taylor unstable plasma segment using the data collected from Solar Terrestrial relations Observatory [48]. Hu et al. theoretically studied the radiation equivalent of RTI-Interface Discontinuous Acceleration phenomena under the incompressible limit of optically thin limit [49]. Ruiz obtained a theoretical model by asymptotically expanding an action principle, which leads to fully nonlinear magnetic Rayleigh–Taylor equations [50]. Jiang and Zhao investigated in a bounded domain of an inhomogeneous incompressible viscous fluid to study the existence of unstable classical solutions of the RT problem [51]. Chen et al. studied the linear RTI of stratified fluids and the effects of elasticity and magnetic fields on it [52]. Jiang et al. reported the phenomenon of inhibition of RTI in the inviscid, inhomogeneous and incompressible fluid by applying horizontal magnetic field along velocity damping [53].

Meshkov and Abarzhi theoretically and experimentally studied the RT flows by focusing of the effect of acceleration on disorder and order in RT mixing [54]. Wang and Zhao mathematically investigated the RTI in the presence of a bounded uniform gravitational field based on Oldroyd-B model inside a compressible viscoelastic fluid [55]. Su uses a sign-changing Taylor sign coefficients for proving the existence of water waves via the method that strong Taylor sign holds in starting time while breaks down at a later time and vice versa [56]. Kord and Capecelatro studied the role of interfacial perturbations to suppress and enhance the RTI growth rate through manipulation of the initial conditions of the evolution of a multi-mode RTI [57]. Li et al. presented a linear analytical model for RTI by considering the long wavelength to study the phase effects in initial confinement fusion caused by the phase difference of single-mode perturbation on two interfaces [58].

Zu et al. worked on the conservative form of Allen–Cahn (A-C) equation for phase field, and they proposed a Lattice Boltzmann (LB) model to track the interface of binary fluid systems [59]. Talat et al. numerically studied RTI problem in 2D by coping with moving boundary conditions and by enabling single domain fixed node approach with the procedure of meshless solution based on diffuse approximate method and phase-field formulation [60]. Liu et al. considered spike for the pure single mode cosine perturbation between interfaces of two classical fluid and employ a method of small parameter expansion for analytical exploration of history of evolution of bubble for potential function having up to third-order nonlinear corrections [61]. Morgan et al. simulate the RTI with large acceleration by studying RTI in gas phase experiments. To make a RTI a stably stratified system of two fluids having a diffused interface separation is accelerated using a rarefaction wave [62]. Wang et al. performed numerical simulations and analytical modeling for the interfacial instability of a water droplet induced by internal volume oscillation [63].

Gebhard and Kolumban considered the evolution of two fluids of homogeneous densities described by the inviscid Boussinesq equations and provides the explicit relaxation of the associated differential inclusion [64]. Bian et al. conducted a systematic analysis on RTI late-time growth for studying the effects of perturbed Atwood number and Reynold number. They show that at large Reynolds number, the bubble re-accelerates [65]. Srinivasan and Hakim work in the nonlinear phase of magneto-RTI and studied the role of electron inertia and Larmor radius effects [66]. Huneault et al. formed gas–liquid interface, which becomes RT unstable, and studied the effect on growth of spray like high mode number perturbation using the implosion of nominally smooth cavity [67]. Terrones and Heberling studied the most-unstable modes by performing a systematic analysis using computational method under the action of a radially directed acceleration over the entire parameter space [68].

Li et al. experimentally studied the evaporation behavior of a silicone-oil-seeded. The observation shows the formation of plumes of segregated fluid and as a result an instantaneous segregation of 1,2-hexanediol in the sessile droplet. [69]. Lherm et al. described the impact catering that crater deceleration after impact, which is responsible for a density-driven perturbation at the drop-pool interface, known as a spherical RTI [70]. Rigon et al. studied the case of young Supernovae Remnants, when interacted with interstellar medium [71]. Polavarapu et al. experimentally presented the result of 3D and 2D perturbation of RTI in the elastic–plastic materials. Results show that by decreasing the initial wavelength and amplitude a more stable interface is produced [72]. Zheng et al. used linear stability analysis to investigate a rigid cylindrical container for RTI confining an elastic soft gel. The gel is found quite sensitive to the aspect ratio and critical load on the surface of the gel [73].

Lyubimova et al. investigated two confined isothermal binary mixture of slowly miscible liquids in horizontal plane layer with the simultaneous convective and diffusive evolution [74]. Awasathi achieved a second-order polynomial in form of dispersion equation for the growth rate analysis of RTI under mass and heat transfer [75]. Chao et al. examined the Spatial-temporal and temporal stability of the system under uniformly cooled–heated inclined substrate and studied the RTI of gravity-driven viscous liquid films flowing [76]. To characterize the instability pattern in spherical geometry, Balestra et al. investigated the RTI of a thin viscous film coated inside a spherical substrate [77]. Livescu et al. used the results from direct numerical simulations of RTI to study the evolution of the flow after the gravity. In turbulent kinetic energy transport equation, the buoyancy term changes sign and becomes a destructive term as the gravity is reversed, which is given by the product between mean pressure gradient and the mass flux, leading to a rapid decay of the kinetic energy [78].

Aslangil et al. explore RTI-induced mixing for zero acceleration with variable periods. The results were compared with the case of classical constant gravity, when acceleration is removed for either some indefinite period or intermediate period [79]. Sabet et al. demonstrated that mixing properties of fluids influenced drastically due to viscosity contrast between two miscible fluids under the destabilization effect of RTI [80]. Chakrabarti et al. addressed the pattern selection of the elastic RTI. The dominant coupling mechanism takes place through three modes of resonance [81]. D'Ortona et al. interested in segregating dry granular flow and studied the self-induced RTI in it [82]. Zhang et al. show that mass ablation can reduce RTI nonlinear bubble growth and RTI is dominated by bubble competition with respect to the classical value for the same initial perturbation amplitude via numerical simulations in 2D and 3D multimode RTI [83].

Yang et al. used a phase-field model to investigate the side wall boundary effect on all boundaries under no-slip conditions. The coupled Cahn–Hillard and Navier–Stokes system equations are solved [84]. Khan and Shah numerically solve the coupled system of incompressible phase-field and Navier–Stokes equations by implementing Neumann and periodic boundary conditions and hence simulated successfully the long-time evolution of the RTI problem [85]. Ding et al. employed molecular dynamic simulation to obtain the growth behavior of RTI [86]. Chen et al. studied the effects of specific heat ratio on the compressible RTI by applying discrete Boltzmann method. They investigate two kind of non-equilibrium quantities related to heat flux and viscous stress [87]. Piriz and Piriz worked on a linear elastic perfect-plastic constitutive model to develop the 2D linear theory of the incompressible RTI [88].

Gancedoet al. studied the dynamics of an incompressible fluid in porous medium driven by capillarity and gravity forces [89]. Zhang et al. worked for spherical geometry and studied the 3D weakly nonlinear RTI by developing a theoretical model for ideal Euler equation and incompressible fluid, and third-order solutions are derived for interface perturbations of spherical harmonic modes [90]. Yang et al. studied the RTI in presence of a horizontal magnetic field along the film for an inviscid flow with finite film thickness. They analyzed the growth of small perturbations on free surfaces [91]. Hillier investigated the physics behind the formation of fragmentation of prominence eruption and plumes in prominences and found that the physical process behind these observations is the magnetic RTI [92]. Dolai andPrajapati investigated the linear RTI in the dust cloud, and hydrodynamics model was formulated by considering Boltzmann distributed electrons and dust particles under magnetic field [93]. Pruss et al. worked on the RTI without and with phase transition in the problem of Verigin [94]. Schulreich and Brietschwerdt demonstrated that powerful time-varying pressure gradient in homogeneous plasma has a sudden energy release effects [95]. Gibbon employed the variable density model and says that RT-turbulence is driven by Navier–Stokes turbulence via gradient of the buoyancy in the inhomogeneous and incompressible [96]. Li et al. theoretically analyzed RTI in gas stream on a spherical droplet [97]. Mahulikar et al. worked on the role of fluctuations that integrate the maximum entropy production and fluctuation theorem in entropy production by a thermodynamic fusion theorem [98].

Zhao. et al. modeled thin shell approximation for an ideal incompressible fluid by extension of thin layer model for nonlinear evolution and deformation of the RTI [99]. Zhou and Cabot utilized three big datasets with a goal of determining degree of departure, using direct numerical simulation with different Atwood numbers at moderate Reynolds number, of homogeneous and isotopic turbulence [100]. Braileanu et al. worked on the development of RTI to understand partial ionization effect between solar corona and magnetized prominence at a smooth realistic transition layer [101]. Zhao et al. generalize thin layer model and put forward thin shell model in 2D spherical geometry for RTI. They obtained equation of motion for the shell analytically, which can also be calculated numerically [102]. Sun et al. work on a unified model, where decomposition of unstable mode into a rotational and irrotational part was done. Non-conservative force and constitutive properties of matter determine the mode of decay with the velocity field [103].

Luttwak used staggered Mesh Godunov scheme to study RTI [104]. Zhang et al. compare the RTI in planar geometry and spherical geometry using the third-order WN solutions for RTI in their research [105]. Zhang et al. using a model established the multi-mode incompressible RTI for spherical geometry in 2D and derived the solutions within the second order. To study the difference between growth of perturbation and geometry effect at the equator and pole, they use the Gaussian-type and cosine-type perturbation [106]. Grouchy et al. experimentally investigated gas-puff Z-pinch implosion and studied Magneto RTI in the cylindrical shocks launched toward the symmetry axis in different plasma of argon and neon [107]. Li et al. developed a code based on Euler method called as massively parallel laser ablative RTI code using simulations by considering the laser energy deposition, hydrodynamics and the electronic thermal conductivity [108]. Kashkovsky et al. performed computation and study using the direct simulation Monte-Carlo method about the development of the RTI and show that it can be successfully simulated at the molecular-kinetic level [109]. The physics of Hall thrusters and existing instabilities are reported in the references [110–113] the Hall thruster support both resistive and RTI. The electromagnetic and electromagnetic oscillation has been discussed in detail in the references [114–119].

3. Introduction to Rayleigh: Taylor instability

Consider a gravitational field in the x direction. If we place a plasma in the field without a vessel, the plasma will fall. Now, consider a situation when a DC magnetic field is applied along $\hat{z}$ ($\vec{B_s} \vee \hat{z}$. The gravitational force on the electrons of plasma can be balanced by the Lorentz force due to magnetic field (**Figure 1**).

This equilibrium, however, is not suitable to perturbation in the plasma boundary. Let the perturbation be denoted by dotted curves. The ion drifts along -Y-direction, and the electrons drift in the positive X-direction. Due to charge separation, an electric field set up in the -Y-direction.

This electric field produces a $\vec{E} \times \vec{B_s}/B_s^2$ drift velocity of electrons and ions in the minus X-direction, and both electrons and ions fall down, which is balanced by the gravitational field. Due to this, perturbation in the boundary grows. This is known as Rayleigh–Taylor instability.

4. Fluid equations and instability analysis

Under this situation, we write the continuity equation

$$\frac{\partial n}{\partial t} + \vec{\nabla} \cdot \left(\vec{v} n\right) = 0 \tag{1}$$

The equation of motion for ion

$$\left(\frac{\partial}{\partial t} + \vec{v} \cdot \vec{\nabla}\right)\vec{v} = \frac{e\left(\vec{E} + \vec{v} \times \vec{B}_0\right)}{M} \tag{2}$$

The equation of motion for electrons

$$\left(\frac{\partial}{\partial t} + \vec{v}_e \cdot \vec{\nabla}\right)\vec{v}_e = -\frac{e\left(\vec{E} + \vec{v}_e \times \vec{B}_0\right)}{m} - \frac{\vec{\nabla} p_e}{\mathrm{mn}_e} \tag{3}$$

4.1 First-order linearization of the equations

We use the linearized form of the above equations for small perturbations density, velocity and electric field. We represent perturbed density by n_1 and velocity by $\vec{v}_1$. The unperturbed density is taken as n_0.The unperturbed value of velocities of ions and electrons is taken as $v_0 = -\frac{\mathrm{Mg}}{\mathrm{eB}_0}\hat{y} = -\frac{g}{\Omega_{\mathrm{ci}}}\hat{y}$ and $v_{0e} = \frac{g}{\Omega_{\mathrm{ce}}}\hat{y}$, respectively. These relative drifts of the plasma species set up a perturbed electric field in the positive Y-direction. Let us write $n = n_0 + n_1$, $\vec{v} = \vec{v}_1 + \vec{v}_0$ and $\vec{E} = \vec{E}_1 + \vec{E}_0$.The linearized form of the above equations thus read

$$\frac{\partial(n + n_1)}{\partial t} + \vec{\nabla} \cdot \left[(n + n_1)\left(\vec{v}_0 + \vec{v}\right)\right] = 0 \tag{4}$$

$$\frac{\partial n_0}{\partial t} + \frac{\partial n_1}{\partial t} + \vec{\nabla} \cdot \left[\left(n_0\vec{v}_0 + \vec{v}_1 n_1 + n_1\vec{v}_0 + n_0\vec{v}_1\right)\right] = 0 \tag{5}$$

$$\frac{\partial n_1}{\partial t} + \left(\vec{v}_0 \cdot \vec{\nabla}\right)n_1 + \left(\vec{v}_1 \cdot \vec{\nabla}\right)n_0 + n_0\left(\vec{\nabla} \cdot \vec{v}_1\right) = 0 \tag{6}$$

The quadratic in perturbation terms $\left(\vec{v}_1 \cdot \vec{\nabla}\right)n_1$. and $n_1\left(\vec{\nabla} \cdot \vec{v}_1\right)$ are neglected. The terms $n_0\left(\vec{\nabla} \cdot \vec{v}_0\right)$ and $n_1\left(\vec{\nabla} \cdot \vec{v}_0\right)$ reduce to zero for the constant velocity $\vec{v}_0$.

Since the mass of the electron is much less than from the mass of the ion, the initial gravitational drift of electrons may be ignored to simply the dispersion relation. Let us consider that ω is the frequency of oscillations and k is the propagation constant of the oscillating quantities. We seek the solution of above equations for which we put $A = A\exp(\mathrm{iky} - i\omega t)$. Replacing $\frac{\partial}{\partial t} \to (-i\omega$ and $\nabla \to \mathrm{ik}$. With the help of Eqs.(5) and (6), we obtain the following expression for the perturbed ion density

$$-i\omega n_{i1} + \mathrm{in}_{i1}kv_0 + v_{\mathrm{ix}}\frac{\partial n_0}{\partial x} + \mathrm{ikn}_0 v_{\mathrm{iy}} = 0 \tag{7}$$

After simplifying, the perturbed density of ions

$$n_{i1} = \frac{\mathrm{kn}_0 v_{\mathrm{iy}} - iv_{\mathrm{ix}}\frac{\partial n_0}{\partial x}}{\omega - kv_0} \tag{8}$$

From the equation of motion of ions, we get

$$\frac{\partial\vec{v}_1}{\partial t} + v_0\frac{\partial\vec{v}_1}{\partial x} = \frac{e}{M}\vec{E}_1 \tag{9}$$

$$\frac{\partial n_{e1}}{\partial t} + u_0\frac{\partial n_{e1}}{\partial y} + n_0\left(\vec{\nabla} \cdot \vec{v}_e\right) + v_{\mathrm{ex}}\frac{\partial n_0}{\partial x} = 0 \tag{10}$$

The linearized form of equations of electrons

$$\frac{\partial \vec{v}_{e1}}{\partial t} + \vec{v}_{0e}\frac{\partial \vec{v}_{e1}}{\partial y} = -\frac{e}{m}\left(\vec{E}_1 + \vec{v}_e \times \vec{B}_0\right) - \frac{\vec{\nabla} p_e}{\mathrm{mn}_0} \tag{11}$$

On the other hand, we obtain from Eq. (11)

$$(\omega - kv_{0e})v_{\mathrm{ex}} = -i\Omega_{\mathrm{ce}}v_{\mathrm{ey}} \tag{12}$$

$$(\omega - kv_{0e})v_{\mathrm{ey}} = -\frac{i\Omega_{\mathrm{ce}}E_y}{B_0} + i\Omega_{\mathrm{ce}}v_{\mathrm{ex}} + \frac{\mathrm{kn}_{e1}T_e}{n_0 m} \tag{13}$$

Here, $\Omega = \frac{\mathrm{eB}_0}{m}$ is taken as the electrongyrofrequency. Above equations are solved for the X- and Y-components under the assumption $\Omega > > \omega - \mathrm{ku}_0$, and we obtain the components

$$v_{\mathrm{ex}} = \frac{E_y}{B_0} - \frac{\mathrm{iV_{th^2}kn}_{e1}}{\Omega_{\mathrm{ce}}n_0} \tag{14}$$

$$v_{\mathrm{ey}} = \frac{i(\omega - kv_{0e})E_y}{\Omega_{\mathrm{ce}}B_0} - \frac{V_{\mathrm{th}^2}\mathrm{kn}_{e1}(\omega - kv_{0e})}{\Omega_{\mathrm{ce}^2}n_0} \tag{15}$$

For ions, their thermal motions velocity has been ignored.

$$v_{\mathrm{ix}} = \frac{E_y}{B_0} \tag{16}$$

And

$$v_{\mathrm{iy}} = -\frac{i(\omega - kv_{0i})E_y}{\Omega_{\mathrm{ci}}B_0} \tag{17}$$

The perturbed ion density

$$n_{i1} = -i\frac{\mathrm{kn}_0E_y}{\Omega_{\mathrm{ci}}B_0} - i\frac{E_y}{B_0(\omega - kv_0)}\frac{\partial n_0}{\partial x} \tag{18}$$

The perturbed electron density can be written as

$$n_{e1} = \frac{v_{\mathrm{ex}}\frac{\partial n_0}{\partial x}}{i(\omega - kv_{e0})} \tag{19}$$

Substituting the value of electron velocity, we get

$$n_{e1} = -\frac{\mathrm{iE}_y\frac{\partial n_0}{\partial x}}{B_0(\omega - kv_{e0})\left\{1 + \frac{V_{\mathrm{th}^2}k\frac{\partial n_0}{\partial x}}{(\omega - kv_{e0})\Omega_{\mathrm{ce}}n_0}\right\}} \tag{20}$$

We use the Poisson's equation with perturbed densities and get

$$\varepsilon_0 \nabla.E = n_{i1} - n_{e1} \tag{21}$$

$$\varepsilon_0 ikE_y = -i\frac{kn_0E_y}{\Omega_{ci}B_0} - i\frac{E_y}{B_0(\omega - kv_0)}\frac{\partial n_0}{\partial x} + \frac{iE_y\frac{\partial n_0}{\partial x}}{B_0(\omega - kv_{e0})\left\{1 + \frac{V_t{}^2k\frac{\partial n_0}{\partial x}}{(\omega - kv_{e0})\Omega_{ce}n_0}\right\}} \tag{22}$$

After solving the above expression, we get the quadratic dispersion equation under the effect of plasma density, wavenumber and gravitation drift of plasma species. To find out the unstable growing waves, we need to find out the unstable eigenvalues of the linearized problem.

$$\omega^2 - (a + b)\omega + \zeta = 0 \tag{23}$$

The various coefficients are given below

$$\begin{aligned} a &= \gamma - c = \gamma - ku_0,\ b = kV_{0i}, \\ c &= ku_0,\ \zeta = ab - \frac{(c - \gamma - b)\beta}{k\alpha}, \\ \alpha &= \epsilon_0 + \frac{n_0}{B_0\Omega_{ci}}, \beta = \frac{n_{0x}}{B_0}, n_{0x} = \frac{\partial n_0}{\partial x}, \\ \gamma &= \frac{kV_{th}^2 n_{0x}}{n_0\Omega_{ce}}. \end{aligned} \tag{24}$$

5. Result and discussion

The above quadratic equation is solved numerically to find out the growth rate and real frequency of Rayleigh–Taylor instability under the effect of thermal motion of electron. It has been shown in **Figure 2** that the growth rate and the real frequency increase linearly with density gradient.

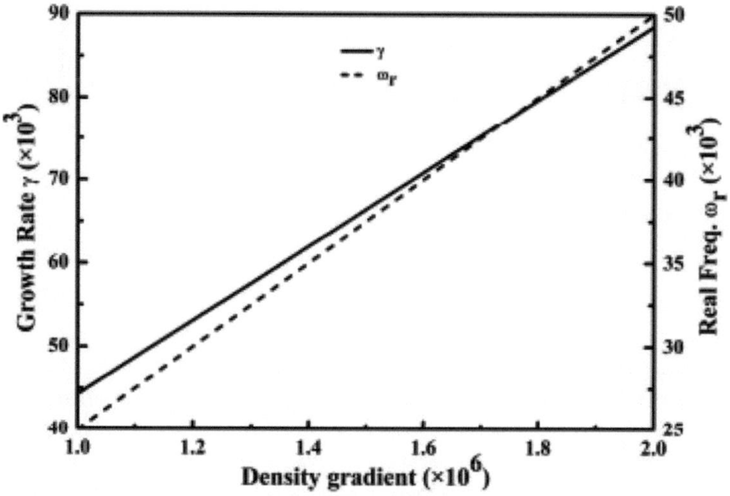

Figure 2.
Variation of growth rate and real frequency with density gradient.

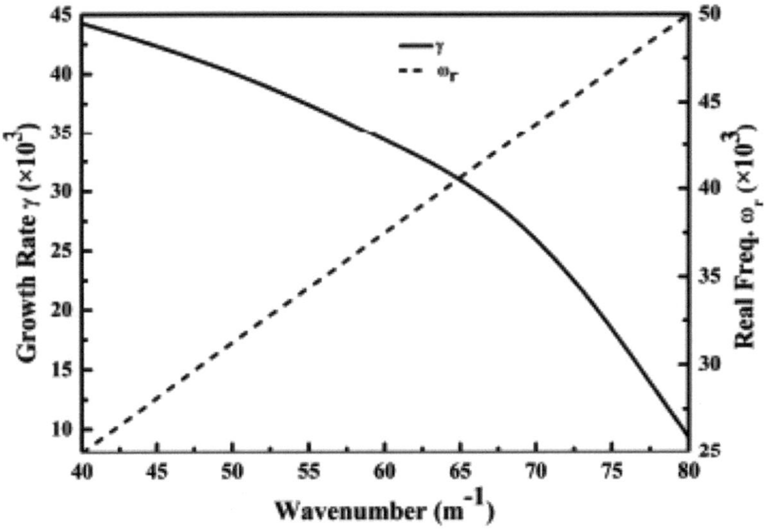

Figure 3.
Variation of growth rate and real frequency with wave number.

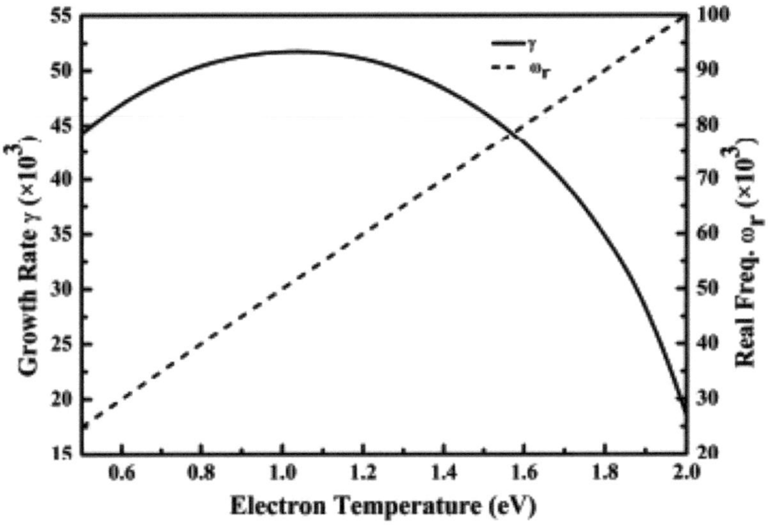

Figure 4.
Variation of growth rate and real frequency with electron temperature.

Figure 3 shows that the growth rate is maximum for the smaller value of wavenumber. In other words, the instability grows faster for the larger wavelength oscillations. In **Figure 4**, the effect of magnetic field on the growth and the real frequency is shown. The growth rate shows almost parabolic variation with magnetic field, and the real frequency decreases with higher value of magnetic field. The real frequency behavior with wavenumber and electron temperature shows that the phase velocity remains constant during the propagation. The growth rate increases with electron temperature up to certain limit, and then, it decreases with higher value of temperature as shown in **Figure 5**.

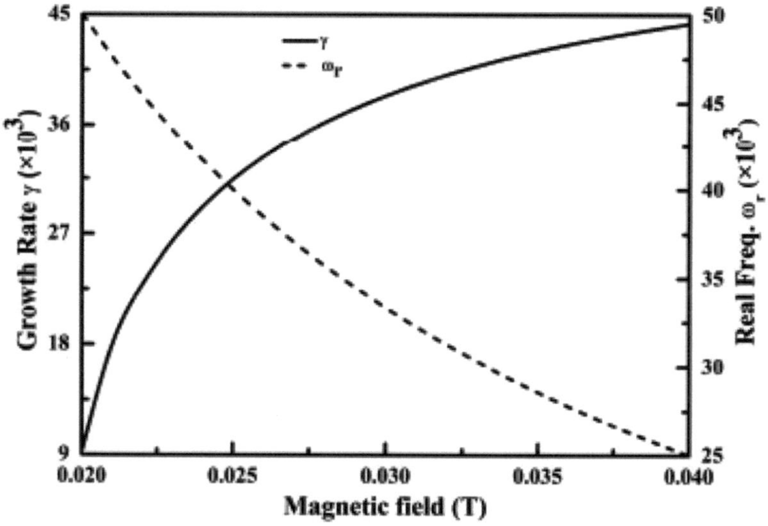

Figure 5.
Variation of growth rate and real frequency with wave number.

6. Conclusions

In Rayleigh–Taylor instability, heavy fluid supported by a light fluid in the presence of relative acceleration, and it measures the rate of mixing of turbulent layers of two fluids of different densities at the interface. The mixing rate and yields can be characterized by measuring the mass and heat exchange across the mixing layers. It is measured that short wavelength perturbations blow up exponentially in RTI. The gravitational force acts on an inverted density gradient and triggers the instability.

Acknowledgements

The University Grants Commission (UGC), New Delhi, India, is thankfully acknowledged for providing the startup Grant (No. F. 30-356/2017/BSR).

Author details

Sukhmander Singh[1*], Bhavna Vidhani[2], Sonia Yogi[2], Ashish Tyagi[3], Sanjeev Kumar[4] and Shravan Kumar Meena[4]

1 Plasma Waves and Electric Propulsion Laboratory, Department of Physics, Central University of Rajasthan, Ajmer, Kishangarh, India

2 Department of Physics and Electronics, Hansraj College, University of Delhi, Delhi, India

3 Physics Department, Swami Shradhan and College, University of Delhi, Delhi, India

4 Department of Physics, Motilal Nehru College University of Delhi, Delhi, India

*Address all correspondence to: sukhmandersingh@curaj.ac.in

References

[1] Chandrasekhar S. Hydrodynamic And hydromagnetic Stability, Dover, 1981, first published by. England: Oxford University Press; 1961

[2] Roberts MS. Fluid instabilities and transition to turbulence. Computational overview of fluid structure interaction. 2020;**4**(23):1-20

[3] Drazin PG. Introduction to Hydrodynamic Stability (Cambridge Texts in Applied Mathematics). United Kingdom: Cambridge University Press; 2002. DOI: 10.1017/CBO9780511809064

[4] Allah MO. An overview of linear and nonlinear Rayleigh-Taylor instability. General Mathematics Notes. 2014;**20**(1): 67-76

[5] Francois C. Hydrodynamic Instabilities. United Kingdom: Cambridge University Press; 2011. DOI: 10.1017/CBO9780511975172

[6] Singh S. In: Ghaedi K et al., editors. Dynamics of Rayleigh-Taylor Instability in Plasma Fluids in Computational Overview of Fluid Structure Interaction, edited by. London: IntechOpen; 2020. DOI: 10.5772/intechopen.92025

[7] Singh S, Malik HK. Role of ionization and electron drift velocity profile to Rayleigh instability in a hall thruster plasma: Cutoff frequency of oscillations. Journal of Applied Physics. 2012; **112**(013307):1-7

[8] Malik HK, Singh S. Conditions and growth rate of Rayleigh instability in a hall thruster under the effect of ion temperature. Physical Review E. 2011; **83**(036406):1-8

[9] Litvak AA, Fisch NJ. Rayleigh instability in hall thrusters. Physics of Plasmas. 2004;**11**:1379-1382

[10] Zanella R, Tegze G, Le Tellier R, Henry H. Two- and three-dimensional simulations of Rayleigh-Taylor instabilities using a coupled Cahn-Hilliard/Navier-stokes model. Physics of fluids. 2020;**32**:124115

[11] Balestra G, Kofman N, Brun P-T, Scheid B, Gallaire F. Three-dimensional Rayleigh-Taylor instability under a unidirectional curved substrate. Journal of Fluid mechanics. 2018;**837**:19-47

[12] Zeng RH, Tao JJ, Sun YB. Three-dimensional viscous Rayleigh-Taylor instability at the cylindrical interface. Physical Review E. 2020;**102**:023112. DOI: 10.1103/PhysRevE.102.023112

[13] Tamim SI, Bostwick JB. A dynamic analysis of the Rayleigh-Taylor instability in soft solids. Extreme Mechanics Letters. 2020;**40**:100940. DOI: 10.1016/j.eml.2020.100940

[14] Gebhard B, Kolumban JJ. A new approach to the Rayleigh–Taylor instability. Archive for rational mechanics and analysis. 2021;**241**: 1243-1280. DOI: 10.1007/s00205-021-01672-1

[15] Zhu G, Shi P, Yang Z, Zheng J, Ying LM, J, Sun X. A new method to suppress the Rayleigh–Taylor instability in a linear device. Physics of Plasmas. 2019;**26**:042107. DOI: 10.1063/1.5087168

[16] Dib YOE, Moatimid GM, Mady AA. A novelty to the nonlinear rotating Rayleigh-Taylor instability. Pramana. 2019;**93**:82. DOI: 10.1007/s12043-019-1844-x

[17] Mondal P, korenaga J. A propagator matrix method for the Rayleigh–Taylor instability of multiple layers: A case study on crustal delamination in the

early earth. Geophysical Journal International. 2018;**212**:1890-1901. DOI: 10.1093/gji/ggx513

[18] Rahman SKM, San O. A relaxation filtering approach for two dimensional Rayleigh-Taylor instability – Induced flows. Fluids. 2019;**4**:78

[19] Song Y, Srinivasan B. A survey of the effects of magnetic field, resistivity, viscosity and thermal conduction on the Rayleigh-Taylor instability. Radiation Effects and Defects in Solids. 2020;**175**: 1009-1014. DOI: 10.1080/ 10420150.2020.1845692

[20] Casner A, Mailliet C, Rigon G, Khan SF, Martinez D, Albetazzi B, et al. From ICF to laboratory astrophysics: Ablative and classical Rayleigh-Taylor instability experiments in turbulent-like regimes. Nuclear Fusion. 2018;**59**:032002

[21] Jha R, Arote A, Banerjee J. Advection Stabilization Using Lower-Order Scheme Blending: A Case Study of Rayleigh–Taylor Instability. In: Prabu T, Viswanathan P, Agrawal A, Banerjee J, editors. Fluid Mechanics and Fluid Power. Lecture Notes in Mechanical Engineering. Singapore: Springer; 2021. DOI: 10.1007/978-981-16-0698-4_88

[22] Yilmaz I. Analysis of Rayleigh-Taylor instability at high Atwood numbers using fully implicit, non-dissipative, energy-conserving large eddy simulation algorithm. Fluids. 2020; **32**:054101. DOI: 10.1063/1.5138978

[23] Ahuja J, Girotra P. Analytical and numerical investigation of Rayleigh-Taylor instability in nanofluids. Pramana. 2021;**95**:1-12. DOI: 10.1007/ s12043-020-02046-0

[24] Zhao Z, Wang P, Liu N, Lu X. Analytical model of nonlinear evolution of single-mode Rayleigh-Taylor instability in cylindrical geometry. Journal of Fluid Mechanics. 2020;**900**: 1-13. DOI: 10.1017/jfm.2020.526

[25] Horne JT, Lawrie AGW. Aspect-ratio-constrained Rayleigh-Taylor instability. Physica D: Nonlinear phenomena. 2020;**406**:132442

[26] Scase MM, Hill RJA. Centrifugally forced Rayleigh-Taylor instability. Journal of fluid mechanics. 2018;**852**: 543-577. DOI: 10.1017/jfm.2018.539

[27] Chen F, Xu A, Zhang G. Collaboration and competition between Richtmyer-Meshkov instability and Rayleigh-Taylor instability. Physics of fluids. 2018;**30**: 102105. DOI: 10.1063/1.5049869

[28] Elyanov A, Golub V, Volodin V. Condition for the development of Rayleigh-Taylor instability on the spherical flame front. Journal of physics: Conference Series. 2018;**1129**:012011. DOI: 10.1088/1742-6596/1129/1/012011

[29] Alqatral S, Thomas EV, Bischofberger I. Confinement-induced stabilization of the Rayleigh-Taylor instability and transition to the unconfined limit. Science advances. 2020;**6**:1-7. DOI: 10.1126/sciadv.abd6605

[30] Gopalakrishnan SS, Landeira JC, Knaepen B, Devit A. Control of Rayleigh-Taylor instability onset time and convective velocity by differential diffusion effects. Physics Review E. 2018;**98**:011101. DOI: 10.1103/ PhysRevE.98.011101

[31] Scase MM, Sengupta S. Cylindrical rotating Rayleigh-Taylor instability. Journal of Fluid Mechanics. 2020;**907**: 1-33. DOI: 10.1017/jfm.2020.842

[32] Sauppe JP, Palaniyappan S, Tobias BJ. Demonstration of scale-invariant Rayleigh-Taylor instability

growth in laser-driven cylindrical implosion experiments. Physical Review Letters. 2020;**124**:185003. DOI: 10.1103/PhysRevLett.124.185003

[33] Shimony A, Malamud G, Shvart D. Density ratio and entrainment effects on asymptotic Rayleigh-Taylor instability. Journal of Fluid Engineering. 2018;**140**: 050906. DOI: 10.1115/1.4038400

[34] Malamud G, Kuranz CC. Design of a single-mode Rayleigh-Taylor instability experiment in the highly nonlinear regime. High Energy Density Physics. 2019;**32**: 18-30. DOI: 10.1016/j.hedp.2019.04.004

[35] Sauppe JP, Malka E. Design of cylindrical implosion experiment to demonstrate scale-invariant Rayleigh-Taylor instability growth. High Energy Density Physics. 2020;**36**:100831. DOI: 10.1016/j.hedp.2020.100831

[36] Wu Q, Shinjo J. Detailed nonlinear dynamics of the liquid spike development in gaseous medium caused by a three-dimensional Rayleigh-Taylor instability. International Journal Of Multiphase flow. 2019;**120**:103107. DOI: 10.1016/j.ijmultiphaseflow.2019.103107

[37] Liang H, Hu X, Huang X, Xu J. Direct numerical simulations of multi-mode immiscible Rayleigh-Taylor instability with high Reynolds numbers. Physics of Fluids. 2019;**31**:112104. DOI: 10.1063/1.5127888

[38] Wolf GGH. Dynamic stabilization of Rayleigh-Taylor instability of miscible liquids and the related frozen waves. Physics of Fluids. 2018;**30**:021701. DOI: 10.1063/1.5017846

[39] Cohen ES, Oron A. Dynamics of nonisothermal two-thin-fluid-layer systems subjected to harmonic tangential forcing under Rayleigh-Taylor instability conditions. Physics of Fluids. 2020;**32**: 082113. DOI: 10.1063/5.0021417

[40] Nimmagadda R, Reddy SR. Dynamics of rising bubbles in gradually mixing fluids due to effect of Rayleigh-Taylor instability. International Journal of Multiphase Flow. 2020;**129**:103288. DOI: 10.1016/j.ijmultiphaseflow.2020.103288

[41] Naveh A, Hill DL, Matthews MT, Abarzhi SI. Early and late-time evolution of Rayleigh-Taylor instability in a finite-sized domain by means of group theory analysis. Fluids Dynamics Research. 2020;**52**:025504

[42] Panda D, Kundu B. Effect of Rayleigh-Taylor instability and ionospheric plasma bubbles on the global navigation satellite system signal. Journal of Asian Earth Sciences. 2019;**170**: 225-233. DOI: 10.1016/j.jseaes.2018.11.006

[43] Hu ZX, Zhang YS, Tian B, He Z, Li L. Effect of viscosity on two dimensional single-mode Rayleigh-Taylor instability during and after the reacceleration stage. Physics of Fluids. 2019;**31**:104108. DOI: 10.1063/1.5122247

[44] Luo T, Wang J, Xie C, Wan M, Chen S. Effects of compressibility and Atwood number on the single-mode Rayleigh-Taylor instability. Physics of fluids. 2020;**32**:012110. DOI: 10.1063/1.5131585

[45] Weiland SA, Hamlington PE, Reckinger SJ, Livescu D. Effects of isothermal stratification strength on vorticity dynamics for single-mode compressible Rayleigh-Taylor instability. Physical review fluids. 2019; **4**:093905. DOI: 10.1103/PhysRevFluids.4.093905

[46] Puillet EJ, Piechaud MR, Brun PT. Elastic amplification of the Rayleigh–

Taylor instability in solidification melts. Proceedings of the National Academy of Sciences. 2021;**118**(10):e2020701118. DOI: 10.1073/pnas.2020701118

[47] Baksht RB, Oreshkin VI, Rousskikh AG, Zhigalin AS. Energy balance in a Z pinch with suppressed Rayleigh–Taylor instability. Plasma Physics and Controlled Fusion. 2018; **60**:035015. DOI: 10.1088/1361-6587/aaa79b

[48] Mishra SK, Singh T, Kashyap P, Srivastava AK. Evolution of magnetic Rayleigh–Taylor instability into the outer solar corona and low interplanetary space. The Astrophysical Journal. 2018;**856**:856. DOI: 10.3847/1538-4357/aaae03

[49] Hu ZX, Zhang YS, Tian BL. Evolution of Rayleigh-Taylor instability under interface discontinuous acceleration induced by radiation. Physical Review E. 2020;**101**:043115. DOI: 10.1103/PhysRevE.101.043115

[50] Ruiz DE. On a variational formulation of the weakly nonlinear magnetic Rayleigh Taylor instability. Physics of Plasmas. 2020;**27**:022121. DOI: 10.1063/1.5132750

[51] Jiang F, Zhao Y. On classical solutions of rayleigh–Taylor instability in inhomogeneous incompressible viscous fluids in bounded domains. ar Xiv preprint ar Xiv. 2019;**1901**(04999):15. DOI: 10.48550/arXiv.1901.04999

[52] Chen Y, Wang W, Zhao Y. On effects of elasticity and magnetic fields in the linear Rayleigh–Taylor instabilityof stratified fluids. Journal of Inequalities and applications. 2018;**203**: 1-31. DOI: 10.1186/s13660-018-1796-6

[53] Jiang F, Jiang S, Zhao Y. On inhibition of Rayleigh–Taylor instability by horizontal magnetic field in an inviscid MHD fluid with velocity damping. Journal of Differential Equations. 2021;**314**:574-652. DOI: 10.1016/j.jde.2022.01.013

[54] Abarzhi SI, Bhowmick AK, Naveh A, Pandian A, Swisher NC, Stellingwerf RF, Arnett WD. Supernova, nuclear synthesis, fluid instabilities, and interfacial mixing. Proceedings of the National Academy of Sciences. 2019;**116** (37):18184-18192

[55] Wang W, Zhao Y. On the Rayleigh–Taylor instability in compressible viscoelastic fluids. Journal of mathematical analysis and application. 2018;**463**:198-221. DOI: 10.1016/j.jmaa.2018.03.018

[56] Su Q. On the transition of the Rayleigh-Taylor instability in 2d water waves. 2020. arXiv preprint arXiv, 1-57. DOI: 10.48550/arXiv.2007.13849

[57] Kord A, Capecelatro J. Optimal perturbations for controlling the growth of a Rayleigh-Taylor instability. Journal of fluid mechanics. 2019;**876**:150-185. DOI: 10.1017/jfm.2019.532

[58] Li ZY, Wang LF, Wu JF, Ye WH. Phase effects of long-wavelength Rayleigh–Taylor instability on the thin Shell. Chinese Physics Letters. 2020;**37**: 025201. DOI: 10.1088/0256-307X/37/2/025201

[59] Zu YQ, Li AD, Wei H. Phase-field lattice Boltzmann model for interface tracking of a binary fluid system based on the Allen-Cahn equation. Physical review E. 2020;**102**:053307. DOI: 10.1103/PhysRevE.102.053307

[60] Talat N, Sarler B. Phase field simulation of Rayleigh–Taylor instability with a meshless method. Engineering Analysis with Boundary Elements. 2018;

87:78-89. DOI: 10.1016/j.enganabound.2017.11.015

[61] Liu W, Wang X, Liu X, Yu C, Fang M, Ye W. Pure single-mode Rayleigh-Taylor instability for arbitrary Atwood numbers. Scientific Reports. 2020;**10**:10. DOI: 10.1038/s41598-020-60207-y

[62] Morgan RV, Cabot WH, Greenough JA, Jacobs JW. Rarefaction-driven Rayleigh-Taylor instability. Part 2. Experiments and simulations in the non-linear regime. Journal of fluid mechanics. 2018;**838**(320):355. DOI: 10.1017/jfm.2017.893

[63] Wang J, Li H, Guo W, Wang Z, Du T, Wang Y, et al. Rayleigh–Taylor instability of cylindrical water droplet induced by laser-produced cavitation bubble. Journal of Fluid Mechanics. 2021;**919**:A42. DOI: 10.1017/jfm.2021.401

[64] Gebhard B, Kolumban JJ. Relaxation of the Boussinesq system and applications to the Rayleigh-Taylor instability. Non-linear Differential Equations and Applications. 2021;**29**: 1-38. DOI: 10.1007/s00030-021-00739-y

[65] Bian X, Livescu D. Revisiting the late-time growth of single-mode Rayleigh-Taylor instability and the role of vorticity. Physica D: Nonlinear Phenomena. 2020;**403**:132250. DOI: 10.1016/j.physd.2019.132250

[66] Srinivasan B, Hakim A. Role of electron inertia and electrons/ion finite Larmor radius effects in low-beta, magneto-Rayleigh-Taylor instability. Physics of Plasmas. 2018;**25**:092108. DOI: 10.1063/1.5046098

[67] Huneault J, Plant D, Higgins AJ. Rotational stabilisation of the Rayleigh-Taylor instability at the inner surface of an imploding liquid shell. Journal of Fluids Mechanics. 2019;**873**:531-567. DOI: 10.1017/jfm.2019.346

[68] Terrones G, Heberling T. Rayleigh–Taylor instability at spherical interfaces between viscous fluids: The fluid/fluid interface. Physics of Fluids. 2020;**32**: 094105. DOI: 10.1063/5.0018601

[69] Li Y, Diddens C, Segers T. Rayleigh–Taylor instability by segregation in an evaporating multicomponent microdroplet. Journal of Fluids Mechanics. 2020;**899**:A22. DOI: 10.1017/jfm.2020.449

[70] Lherm V, Deguen R, Alboussiere T, Landeau M. Rayleigh–Taylor instability during impact cratering experiments. Journal of Fluid Mechanics. 2022;**937**:A20

[71] Rigon G. Rayleigh-Taylor instability experiments on the LULI2000 laser in scaled conditions for young supernova remnants. Physical Review E. 2019;**100**: 021201. DOI: 10.1103/PhysRevE.100.021201

[72] Polavarapu R, Roach P, Banerjee A. Rayleigh-Taylor-instability experiments with elastic-plastic materials. Physical Review E. 2019;**99**:053104. DOI: 10.1103/PhysRevE.99.053104

[73] Zheng Y, Cai S. Rayleigh-Taylor instability in a confined elastic soft cylinder. Journal of the Mechanics and Physics of Solids. 2019;**131**:221-229. DOI: 10.1016/j.jmps.2019.07.006

[74] Lyubimova T. Rayleigh-Taylor instability of a miscible interface in a confined domain. Physics of Plasmas. 2019;**31**:014104. DOI: 10.1063/1.5064547

[75] Awasthi MK. Rayleigh-Taylor instability of swirling annular layer with mass transfer. Journal of Fluids Engineering. 2019;**141**:07122. DOI: 10.1115/1.4042174

[76] Chao Y, Zhu L, Yuan H. Rayleigh-Taylor instability of viscous liquid films under a temperature-controlled inclined substrate. Physical Review Fluids. 2021; **6**:064001. DOI: 10.1103/PhysRevFluids.6.064001

[77] Balestra G, Nguyen DMP, Gallaire F. Rayleigh-Taylor instability under a spherical substrate. Physical Review Fluids. 2018;**3**:084005. DOI: 10.1103/PhysRevFluids.3.084005

[78] Livescu D, Brady PT. Rayleigh–Taylor instability with gravity reversal. Physica D: nonlinear phenomena. 2021; **417**:132832. DOI: 10.1016/j.physd.2020.132832

[79] Aslangil D, Farley Z, Lawrie AGW, Banerjee A. Rayleigh–Taylor instability with varying periods of zero acceleration. Journal of Fluids Engineering. 2020;**142**:121103. DOI: 10.1115/1.4048348

[80] Sabet N, Hassanzadeh H, Wit AD, Aabedi J. Scalings of Rayleigh-Taylor instability at large viscosity contrasts in porous media. Physical Review Letters. 2021;**126**:094501. DOI: 10.1103/PhysRevLett.126.094501

[81] Chakrabarti A, Audoly B. Selection of hexagonal buckling patterns by the elastic Rayleigh-Taylor instability. Journal of the Mechanics and Physics of Solids. 2018;**121**:234-257. DOI: 10.1016/j.jmps.2018.07.024

[82] D'Ortona U, Thomas N. Self-induced Rayleigh-Taylor instability in segregating dry granular flows. Physical Review Letters. 2020;**124**:178001. DOI: 10.1103/PhysRevLett.124.178001

[83] Zhang H, Betti R, Yan R, Zhao D, Shvarts D, Aluie H. Self-similar multimode bubble-front evolution of the ablative Rayleigh-Taylor instability in two and three dimensions. Physical Review Letters. 2018;**121**:185002. DOI: 10.1103/PhysRevLett.121.185002

[84] Yang J, Kim J. Side wall boundary effect on the Rayleigh–Taylor instability. European Journal of Mechanics- B/Fluids. 2021;**85**:361-374. DOI: 10.1016/j.euromechflu.2020.10.001

[85] Khan SA, Shah A. Simulation of the two-dimensional Rayleigh-Taylor instability problem by using diffuse-interface model. AIP Advances. 2019;**9**: 085312. DOI: 10.1063/1.5100791

[86] Ding J, Sun P, Huang S, Luo X. Single- and dual-mode Rayleigh–Taylor instability at microscopic scale. Physics of Fluids. 2021;**33**:042102. DOI: 10.1063/5.0042505

[87] Chen L, Lai H, Lin C, Li D. Specific heat ratio effects of compressible Rayleigh-Taylor instability studied by discrete Boltzmann method. Frontiers of Physics. 2021;**16**:52500

[88] Piriz AR, Piriz SA, Tahir NA. Stability boundaries for the Rayleigh-Taylor instability in accelerated elastic-plastic solid slabs. Physical Review E. 2019;**100**:063104. DOI: 10.1103/PhysRevE.100.063104

[89] Gancedo F, Scrobogna S. Surface tension stabilization of the Rayleigh-Taylor instability for afluid layer in a porous medium. Annales de l'Institut Henri Poincaré C, Analyse non linéaire. 2020;**37**:1299-1343. DOI: 10.1016/j.anihpc.2020.04.005

[90] Zhang J, Wang JF, Wu JF, Ye WH, Zou SY, Ding YK, et al. The three-dimensional weakly nonlinear Rayleigh–Taylor instability in spherical geometry. Physics of Plasmas. 2020;**27**:022707. DOI: 10.1063/1.5128644

[91] Yang F, Stone HA. The effects of a horizontal magnetic field on the Rayleigh–Taylor instability. Nuclear Materials and Energy. 2019;**18**:175-181. DOI: 10.1016/j.nme.2018.12.023

[92] Hillier A. The magnetic Rayleigh–Taylor instability in solar prominences. Reviews of Modern Plasma Physics. 2018; **2**:1-47. DOI: 10.1007/s41614-017-0013-2

[93] Dolai B, Prajapati RP. The rotating Rayleigh-Taylor instability in a strongly. Physics of Plasmas. 2018;**25**:083708. DOI: 10.1063/1.5041088

[94] Pruss J, Simonett G, Wilke M. The rayleigh-Taylor instability for the Verigin problem with and without phase transition. Nonlinear Differential Equations and Applications. 2019;**26**: 1-35. DOI: 10.1007/s00030-019-0564-8

[95] Schulreich MM, Breitschwerdt D. The time-dependent Rayleigh–Taylor instability in interstellar shells and supershells, including the eROSITA bubbles. Monthly Notices of the Royal Astronomical Society. 2022;**509**:716-737. DOI: 10.1093/mnras/stab2940

[96] Gibbon JD. The variable density model for the Rayleigh-Taylor instability and its transformation to the diffusive, inhomogeneous, incompressible Navier-stokes equations. Physical Review Fluids. 2021;**6**:1-7. DOI: 10.1103/PhysRevFluids.6.L082601

[97] Li Y, Kang N. Theoretical analysis of Rayleigh-Taylor instability on a spherical droplet in a gas stream. Applied Mathematical Modelling. 2019;**67**: 634-644. DOI: 10.1016/j.apm.2018.11.046

[98] Mahulikar SP, Sengupta TK, Sharma N, Rastogi P. Thermodynamic merger of fluctuation theorem and principle of least action: Case of Rayleigh–Taylor instability. Journal of Non-Equilibrium Thermodynamics. 2019;**44**:363-371. DOI: 10.1515/jnet-2018-0091

[99] Zhao KG, Wang LF, Xue C, Ye WH, Wu JF, Ding YK, et al. Thin layer model for nonlinear evolution of the Rayleigh-Taylor instability. Physics of Plasmas. 2018;**25**:032708. DOI: 10.1063/1.5009257

[100] Zhou Y, Cabot WH. Time-dependent study of anisotropy in Rayleigh-Taylor instability induced turbulent flows with a variety of density ratios. Physics of Fluids. 2019;**31**:084106. DOI: 10.1063/1.5110914

[101] Braileanu BP, Lukin VS, Khomenko E, Vicente AD. Two-fluid simulations of Rayleigh-Taylor instability in a magnetized solar prominence thread. Astronomy and Astrophysics. 2021;**646**:1-12. DOI: 10.1051/0004-6361/202039053

[102] Zhao KG, Xue C, Wang LF. Two-dimensional thin shell model for the nonlinear Rayleigh-Taylor instability in spherical geometry. Physics of Plasmas. 2019;**26**:022710. DOI: 10.1063/1.5079316

[103] Sun YB, Tao JJ, He XT. Unified decomposition method to study Rayleigh-Taylor instability in liquids and solids. Physical Review E. 2018;**97**: 063109. DOI: 10.1103/PhysRevE.97.063109

[104] Luttwak G. Using the SMG scheme to study the Rayleigh-Taylor instability growth in solids. Computers and Fluids. 2020;**208**:104603. DOI: 10.1016/j.compfluid.2020.104603

[105] Zhang J, Wang JF, Ye WH. Weakly nonlinear incompressible Rayleigh-Taylor instability in spherical and planar

geometries. Physics of Plasmas. 2018;**25**: 022701. DOI: 10.1063/1.5017749

[106] Zhang J, Wang LF, Ye WH. Weakly nonlinear multi-mode Rayleigh-Taylor instability in two-dimensional spherical geometry. Physics of Plasmas. 2018;**25**: 082713. DOI: 10.1063/1.5043529

[107] Grouchy PWL, Kusse BR, Banasek J. Observations of the magneto-Rayleigh-Taylor instability and shock dynamics in gas-puff Z-pinch experiments. Physics of Plasmas. 2018; **25**:072701. DOI: 10.1063/1.5032084

[108] Li Z, Wang L, Wu J, Ye W. Numerical study on the laser ablative Rayleigh–Taylor instability. Acta Mech. Sin. 2020;**36**:789-796. DOI: 10.1007/s10409-020-00933-8

[109] Kashkovsky AV, Kudryavstsev AN, Shershnev AA. Numerical simulation of the Rayleigh-Taylor instability in rarefied Ar/He mixture using the direct simulation Monte Carlo method. Journal of Physics: Conference Series. 2019;**1382**: 012154. DOI: 10.1088/1742-6596/1382/1/012154

[110] Singh S. In: Shahzad A, editor. Waves and Instabilities in E × B Dusty Plasma in Thermophysical Properties of Complex Materials. London: IntechOpen; 2019. DOI: 10.5772/intechopen.90397

[111] Singh S. In: Ghaedi K et al., editors. Evolutions of Growing Waves in Complex Plasma Medium in Computational Overview of Fluid Structure Interaction, edited by. London: IntechOpen; 2020. DOI: 10.5772/intechopen.93232

[112] Singh S et al. . In: Matsuuchi K, Hasegawa H, editors. Introduction to Plasma Based Propulsion System: Hall Thrusters in Propulsion: New Perspectives and Applications. London: IntechOpen; 2021. DOI: 10.5772/intechopen.96916

[113] Singh S. In: Singh S, editor. Hall Thruster: An Electric Propulsion through Plasmas in Selected Topics in Plasma Physics. London: IntechOpen; 2020. DOI: 10.5772/intechopen.91622

[114] Singh S, Tyagi A, Vidhani B. In: Wang L, editor. Physics of Absorption and Generation of Electromagnetic Radiationin Electromagnetic Wave Propagation for Industry and Biomedical Applications. London: IntechOpen; 2021. DOI: 10.5772/intechopen.99037

[115] Singh S, Vidhani B, Tyagi A. In: Shahzad A, editor. Numerical Investigations of Electromagnetic Oscillations and Turbulences in Hall Thrusters Using Two Fluid Approach in Plasma Science and Technology. London: IntechOpen; 2021. DOI: 10.5772/intechopen.99883

[116] Singh S, Malik HK, Nishida Y. High frequency electromagnetic resistive instability in a hall thruster under the effect of ionization. Physics of Plasmas. 2013;**20**:102109

[117] Singh S, Malik HK. Growth of low frequency electrostatic and electromagnetic instabilities in a hall thruster. IEEE Transactions on Plasma Science. 2011;**39**:1910-1918

[118] Singh S, Malik HK. Resistive instabilities in a hall thruster under the presence of collisions and thermal motion of electrons. The Open Plasma Physics Journal. 2011;**4**:16-23. DOI: 10.2174/1876534301104010016

[119] Malik HK, Singh S. Resistive instability in a hall plasma discharge under ionization effect. Physics of Plasmas. 2013;**20**:052115

Section 2

Nonlinear Phenomena in Plasma

Chapter 3

Various Aspects of Dust-Acoustic Solitary Waves (DAWs) in Inhomogeneous Plasmas

Hirak Jyoti Dehingia

Abstract

Dusty plasma offers an extensive study of space and astrophysical environments. In this chapter, we have studied some of the basic properties of dusty plasmas, interaction of dust and plasma particles, and the effect of intramolecular attraction and repulsion between these plasma and dust grain particles. During these investigations, we have discussed a few basic relations of dusty plasma and the effect of dust particles on the nonlinear wave structures in dusty plasma. Here, we have also studied the various aspects of dust-acoustic solitary waves (DASWs) in inhomogeneous plasma. To study and analyze the various aspects of DAWs in inhomogeneous plasmas, the governing fluid equations of plasmas are considered to derive the *Korteweg de-Vries (KdV)* equation. The solution of the *KdV* equation is obtained as soliton or solitary wave. The solitary wave solution indicates the various characteristics of DASWs in the inhomogeneous dusty plasma. In this chapter, a systematic and extensive study on DAWs is also included for the inhomogeneous and unmagnetized plasmas.

Keywords: dusty plasma, inhomogeneous plasmas, soliton, dust-acoustic waves (DAWs), dust-plasma interaction

1. Introduction

Plasmas and dust particles are the two fundamental elements of the universe. The interaction of these two basic ingredients forms a new area of the research field called dusty plasma or complex plasma. Dusty plasma is a complex plasma system consisting of plasma and charged dust particles. Dusty plasmas are abundant everywhere in the universe. Many examples of dusty plasma can be seen in the visible universe—Saturn rings, interstellar clouds and circumstellar clouds, comets, and Mars rings [1]. In the present days of plasma research, dusty plasma is also one of the advanced and trending research fields. As an application of dusty plasma, it includes the study of geophysics, space missions, advancement in astrophysical areas such as a better understanding of the formation of galaxies and stars or planets, and applications in the industries for producing modern materials. Besides the astrophysical environments, dusty plasma is also considered one of the essential fields to study and work in controlled thermonuclear research (CTR). In 1982, the Voyager spacecraft [2] discovered some interesting facts about Saturn's B ring and the radial spokes for the first time. Including the astrophysical issues, dusty plasma has excellent applications in manufacturing new materials in semiconductor industries [3].

Many researchers have done both theoretical and experimental works in dusty plasma. To understand the fundamental properties of duty plasma, some devices, such as the rotating drum [4] and dust shaker systems [5], are newly introduced in the laboratory. Researchers like Northrop [6] and Goertz [7] have added some theoretical works in the field of the dusty plasma system. In unmagnetized dusty plasmas, usually, there are three kinds of normal modes of waves considered. They are dust-acoustic waves (DAW) [8], dust ion-acoustic waves (DIAW) [9], and dust-lattice waves (DLW) [10]. Some theoretical and empirical studies [11, 12] have been done to understand the linear behaviors of dusty plasma in various physical situations. Some of the researchers have done their work on the instabilities of nonlinear waves in dusty plasmas [13–20]. For studying the linear wave theories for small amplitude waves in plasma physics, the nonlinearities are omitted, but the nonlinearities cannot be omitted for the high-amplitude waves. The existence of nonlinearity indicates some novel behaviors of the nonlinear wave structures, such as solitons, supersolitons, rouge waves, and shock waves. Theoretically, the various properties of dust lattice solitary waves [10, 21], DA solitary waves [9, 22, 23] DIA solitary waves [24, 25], and shock waves [26–29] have been studied in an inhomogeneous plasma. Complex plasmas have extensive applications in astrophysical systems, space science, and laboratory experiments [30, 31]. Atteya et al. studied the properties of DASWs consisting of negatively and positively charged dust grains in the presence of κ - distributed superthermal electrons [32]. Earlier, many research works have been done based on homogeneous plasma systems. Still, in the present days, inhomogeneous plasma systems are also widely discussed to understand the various applications of dusty plasma in space and laboratory discharges [33]. The dust-cyclotron wave and modulational instability of dust-acoustic solitary waves have been investigated in the three-component plasma for electrons, ions, and dust particles in a magnetic field [34]. Recently, a study has been done on magneto-acoustic waves propagating and pair-ion fullerene plasma in linear and nonlinear environments [35]. The dust grains are negatively charged and unstable due to their concentration of weights [36]. With the increase in the electron density, the phase velocity of the dust ion-acoustic solitary waves decreases [37, 38]. The DIASWs were also studied both theoretically and experimentally. The DIASW was studied in the presence of a collisionless plasma [39] for some limiting cases, but it was not well designed in some real conditions [40]. Ghosh et al. [41] investigated the effect of damping on DIAWs. Losseva et al. [42] studied the evolution of weakly dissipative hybrid DIASW in the presence of collisional plasma. Popel et al. [43] experimentally investigated the wave velocity, which is much larger than the wave velocity in theoretical conditions. Kruskal [44] has also observed the gravitational wave for inviscid fluid that can asymptotically describe the Korteweg-de Vries (KdV) equation. Some researchers have studied solitary waves, DIASWs, and shock waves both theoretically and experimentally [11, 45–47] in dusty plasmas. Ashraf et al. [48] observed the characteristics of obliquely propagating DIASWs in the presence of nonextensive dusty plasma. Bacha et al. [49] studied the DIASW in the presence of non-Maxwellian dusty plasmas. Tribeche and Zerguini [50] studied the effect of modulations and instabilities in DIASWs in the presence of collisional dusty plasma. Later, an investigation revealed that a plasma system consisting of Gaussian-distributed charged dust grains exhibits DIASWs properties in the presence of a collisional effect [51].

Similarly, a few works have been established [52–55] on the solution of the damped KdV equation in various plasma systems using the classical fluid approach of plasma physics. The application of plasma particles is observed under the effect of nonextensive electron distributions in astrophysical and cosmological scenarios [56], plasma dynamics [57], nonlinear gravitational model [58], and Hamiltonian systems [59] in the presence of long-range interaction. The impact of nonextensive plasma particles on various waves propagating in inhomogeneous plasma is studied in works [60–63]. An extensive study on the distribution of nonextensive electrons is seen in the results [64].

This chapter discusses some essential characteristics of dusty plasma and the various aspects of the DASWs in the inhomogeneous plasma system. We have presented our study and analysis of the DASWs in inhomogeneous plasmas. During the investigations, we considered the governing fluid equations to derive the Korteweg-de Vries (KdV) equation. The characteristic behaviors of DASWs and the interaction of dust and plasma particles are also studied in various physical situations.

2. Properties of dusty plasmas

The dust particles consist of some fundamental elements, such as electrons, ions, and charged dust particles. In a dusty plasma system, there are three characteristic scale lengths of the complex plasma system. They are the average intergrain distance, the Debye radius, and the dust grain radius. In a dusty plasma system, the Debye radius λ_D is given by [65]

$$\frac{1}{\lambda^2{}_D} = \frac{1}{\lambda^2{}_{De}} + \frac{1}{\lambda^2{}_{Di}} \tag{1}$$

where λ_{Di} and λ_{De} are the ion Debye radius and electron Debye radius of a dusty plasma system, respectively. In a dusty plasma mixture, the dust grain particles are one of the most fundamental and essential elements. These dust grain particles are comparatively larger than the negatively charged electrons or positively charged ions in the plasma medium. In a dusty plasma system, the quasi-neutrality condition for the negatively charged dust grains is given by

$$n_{i0} = n_{e0} + Z_{d0}n_{d0} \tag{2}$$

where Z_{d0} is the unperturbed number density of charged dust grains and n_{k0} is the unperturbed number density of the particle species k. Here, the particle species k stands for species of electrons, ions, and dust gains, respectively (I = ions, e = electrons, and d = dust grains). When the dust grains and electrons are attached, we have the following equation:

$$n_{e0} \ll Z_{d0}n_{d0} \tag{3}$$

However, due to the presence of a minimum value ratio of the number densities of ions to the electrons, the reduction of electrons cannot be done completely. While the surface potential of the dust particles tends to zero, the ratio of the number densities of the ions and electrons is not seen to have vanished. Usually, these types of situations, that is, depletion of electrons occur in the Saturn rings, as well as in the laboratory discharges. However, in the thermal complex plasmas, the dust particles get positively charged by releasing electrons. So, by omitting the electrons from charged dust particles, the dust particles become positively charged. Therefore, at an equilibrium condition, we get the following equation:

$$n_{e0} \approx Z_{d0} n_{d0} \tag{4}$$

The Coulomb coupling parameter for the strongly correlated dust particles in a dusty medium is given by

$$\Gamma = \frac{q^2}{dT_d} \exp(-\kappa) \tag{5}$$

where, $q = Z_{d0}e$ is the unperturbed number density of dust particles, $\kappa = \frac{d}{\lambda_D}$,and T_d is the temperature of dust grains. In the case of weakly coupled dusty plasma, we have $\Gamma \ll 1$, and for strongly coupled dusty plasma, it is found to be $\kappa \ll 1$ and $\Gamma \gg 1$. To understand the dynamics and crystal formations in dusty plasma, the strongly coupled dusty plasmas are kept along in the low-temperature laboratory discharges in strongly coupled dusty plasmas. In strongly coupled dusty plasmas, the colloidal systems and laser implosion experiments both show a strong linkage between them.

3. Interactions of dust and plasma particles

The size of the dust particles or dust grains differs from the other plasma particles, that is, from ions to electrons. The dust particles are very large compared to these plasma particles. They are dependent on their surroundings. The size of the dust charges depends on the inclusion of ions, electron fluxes, and the local parameters, but the plasma particles are depending on the distribution functions. The fluxes are influenced by the inclusion of dust particles, so there is also a change in the charges of dust grains. Thus, the changes in the interaction of dust particles with other plasma particles are seen accordingly.

The existence of new forces between the plasma particles and dust grains can also be experienced even if there are fixed charges. In this case, the interaction between the dust grain particles and plasma particles could be different. The interactions among the dust particles could vary due to the variations in the distance of the dust particles and Debye length. These can be experimented by Etching experiments and plasma-dust crystal experiments. The distance between the dust particles and the Debye length is very less in the Etching experiment, but on the other hand, the distance between the dust particles and the Debye length is very large in plasma-dust crystal experiment. In plasma-dust crystal experiments, the Coulomb interaction and mutual shadowing of plasma fluxes both are given much importance.

4. Effect of intramolecular dust particle attraction and repulsion

The nonlinear structures of dusty plasma can redistribute the dust particles in their structures. We may assume that the dust particles will be concentrated in a single place in the nonlinear structure. Thus, it will result in the dust particle as negatively charged dust grains due to the fluctuations of electrons and ions. In the case of dusty plasma, there is no interaction of dust particles with neutral particles. Then the structures of the nonlinear DA waves will be modified by the dust particles that are to be drift vortices. If there is an interaction between the dust particles and neutral gas particles, there will be a strong bond of vortices. It will exhibit the simultaneous drift and gas vortices of the nonlinear wave structures.

In the upper atmosphere and the lower ionosphere, there may exist some nonlinear wave structures due to the existence of gas and drift vortices. Due to the existence of a degree of pollution and dust particles in the upper atmosphere, some motion can be observed in the linkage to the usual vortex motion. The solar activity and degree of ionization influence the drift structures of the nonlinear waves in the inhomogeneous duty plasma systems.

5. Basic relations in dusty plasma

1. *Debye length* (λ_D) *and dust size* (D): In the case of smaller dust particles, that is, $D \ll \lambda_D$, we have obtained the Eq. (1) as follows:

$$\frac{1}{\lambda^2_D} = \frac{1}{\lambda^2_{De}} + \frac{1}{\lambda^2_{Di}}. \tag{6}$$

Then for $D = \lambda_{Di}$ and $T_e \gg T_i$, we have obtained the dimensionless equation for dust size D as follows:

$$D = \frac{D_0}{\lambda_{Di}} \tag{7}$$

2. *Charging of a dust particle:* Floating potential argument shows an assumption on the charge of the dust particle in the units of electron charge e is given by $\frac{Z_d e^2}{D} \sim T_e$. Then the dimensionless equation for the dust charge z is given by

$$z \equiv \frac{Z_d e^2}{DT_e} \tag{8}$$

3. *Dimensionless dust density:* In the dust-plasma crystal experiment, the dimensionless dust density ρ is of order 1; in the etching experiment, it is of order 10; and in the astrophysical situations, it varies from 10 to 10^{-3}, which is shown by the equation given by

$$\rho \equiv \frac{n_d Z_d}{n_e} \tag{9}$$

where n_d is the normalized number density of dust particles. The quasi-neutrality condition for dusty plasma is given by

$$n_{i,0} = n_{e,0}(1 + \rho) \tag{10}$$

6. Methodology to solve nonlinear equations

Nonlinear plasma waves follow governing fluid equations of plasma in various physical circumstances. Due to the existence of nonlinear plasma inhomogeneity, plasma is considered to be one of the frontier research areas in science and engineering. Using fluid approach in plasma physics, lots of research has been carried out for understanding the soliton propagation, soliton reflection, sheath and beam formation, etc. For the last many decades, researchers are using one of the major techniques (method) for deriving and solving nonlinear waves in plasma physics, which is *reductive perturbation technique (RPT).* This *RPT* is very useful for the nonlinear small amplitude waves. In this technique, both the space and time coordinates are transformed to new stretched coordinates for obtaining the family of nonlinear equations, such as *Korteweg-de Vries (KdV) equation, nonlinear Schrödinger equation (NLSE),* and *Kadomtsev Petviashvili (KP)* equation.

For dust-acoustic waves (DAWs), the governing fluid equations are considered as follows:

Continuity equation:

$$\frac{\partial n_d}{\partial t} + \frac{\partial (n_d v_d)}{\partial x} = 0 \tag{11}$$

Momentum equation:

$$\frac{\partial v_d}{\partial t} + n_d \frac{\partial v_d}{\partial x} + q \frac{\partial \varnothing}{\partial x} = 0 \tag{12}$$

Electron distribution equation:

$$n_e = \exp(\phi) \tag{13}$$

Ion distribution equation:

$$n_i = \exp(-\phi) \tag{14}$$

Poison equation:

$$\frac{\partial^2 \phi}{\partial^2 x} = n_e - n_i + q n_d \tag{15}$$

In this technique, we have considered a set of dependent variables in terms of power of smallness parameter $\boldsymbol{\varepsilon}$. Then using the set of stretched coordinates [66] $\xi = \varepsilon^{1/2}(x - \lambda t), \tau = \varepsilon^{3/2} t$, we get the k-dv equation as follows:

$$\frac{\partial \phi_1}{\partial \tau} + A \phi_1 \frac{\partial \phi_1}{\partial \xi} + B \frac{\partial^3 \phi_1}{\partial \xi^3} = 0 \tag{16}$$

In the above *K-dV* equation, the constants $\boldsymbol{A}$ and $\boldsymbol{B}$ represent the nonlinearity and dispersive effect, respectively, in a considered plasma system. After solving the above equation and using some particular transformations, we get the required soliton solution $\boldsymbol{\phi_1}$.

7. Effects of dust particles on the nonlinear wave structures

Dust acoustic (DA) solitary waves: The dimensionless nonlinear governing fluid equations for the DA waves in inhomogeneous unmagnetized plasmas are given by [9]

$$\frac{\partial n_d}{\partial t} + \frac{\partial (n_d u_d)}{\partial x} = 0 \tag{17}$$

$$\frac{\partial u_d}{\partial t} + u_d \frac{\partial u_d}{\partial x} = -\frac{\partial \varnothing}{\partial x} \tag{18}$$

$$\frac{\partial^2 \phi}{\partial x^2} = n_d + \gamma_e e^{\sigma_i \phi} - \gamma_i e^{-\phi} \tag{19}$$

where u_d is the normalized fluid velocity of dust grains, $\gamma = {}^{n_{e0}}/_{n_{i0}}$, and ϕ is the normalized electrostatic wave potential. Here, x and t are the space variable and time variables normalized by λ_D and ion plasma period (ω), respectively. We have considered that the electrons and ions are distributed in the *Boltzmannean distribution* functions in accordance. We have also considered $\gamma_e = \gamma/(1-\gamma)$, $\sigma_i = T_i/T_e$, and $\gamma_i = 1/(1-\gamma)$.

To employ the reductive perturbation technique (RPT), the stretched coordinates [66] are chosen as $\xi = \epsilon^{\frac{1}{2}}(z - Mt)$ and $\tau = \epsilon^{3/2}t$, where ϵ is a smallness expansion parameter and M is the normalized phase velocity of the soliton. The expanded dependent variables n_d, v_d, Z_d, and ϕ in terms of power series of ϵ are given by:

$$\begin{bmatrix} n_d \\ v_d \\ \phi \\ Z_d \end{bmatrix} = \begin{bmatrix} n_{d0} \\ 0 \\ 0 \\ Z_{d0} \end{bmatrix} + \epsilon \begin{bmatrix} n_{d1} \\ v_{d1} \\ \phi_1 \\ Z_{d1} \end{bmatrix} + \epsilon^2 \begin{bmatrix} n_{d2} \\ v_{d2} \\ \phi_2 \\ Z_{d2} \end{bmatrix} + \dots \tag{20}$$

Now, using (20) and RPT on the Eqs. (17)–(19), we have obtained the *Korteweg-de Vries (KdV)* equation, which is of the form

$$\frac{\partial \phi^{(1)}}{\partial \tau} + U \frac{\partial \phi^{(1)}}{\partial \xi} + V \frac{\partial^3 \phi^{(1)}}{\partial \xi^3} = 0 \tag{21}$$

where $U = -\frac{M^3}{(1-\gamma)^2}[1 + (3 + \sigma_i \gamma)\gamma\sigma_i + 0.5\gamma(1 + \sigma_i^2)]$ and $V = 0.5M^3$.

8. Results and discussions

Solving the Eq. (21), we get the solitary wave solution or soliton w.r.t. The velocity v is given by [37]

$$\phi^{(1)} = \phi_m \, sec\, h^2 \left[\frac{U}{W}\right] \tag{22}$$

where $P = \xi - v\tau$, $A = 0.5\gamma^{-1.5}$, and v is the normalized constant velocity, $\phi_m = \frac{3v}{c}$ and $W = \sqrt{\frac{4A}{v}}$ are amplitude and width of the DASWs, respectively. As $M > 0$ and Eq. (22) indicates and we always get the result $c < 0$ for every possible value of σ_i and γ. Thus, Eq. (22) also indicates that the dust acoustic (DA) solitary waves are also included in dusty plasma for the negative potential.

From the above **Figure 1**, we have observed that the amplitude of dust-acoustic (DA) solitary wave decreases with the increase in the number of dust grains. This implies that if the density of dust particles is compact in a particular region of space in the

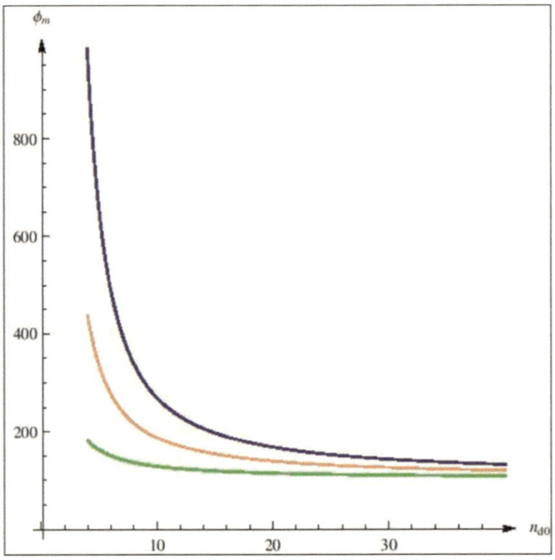

Figure 1.
Variations in amplitude of DASW w.r.t. number density of dust grains (n_{do}).

plasma system, the soliton will propagate smoothly with a lower amplitude wave. In this case, the solitary wave will show fewer structural changes propagating through the inhomogeneous plasma medium. But in the case of **Figure 2**, we have seen that as the number density of dust particles increases, the width DA solitary wave increases. This result indicates that the solitary wave will propagate with a larger area than the soliton amplitude. The solitary wave will propagate slowly throughout the plasma medium during this propagation. When the electron and ion distribution is relatively high with the higher dust number density, the phase of the solitary wave gets boosted.

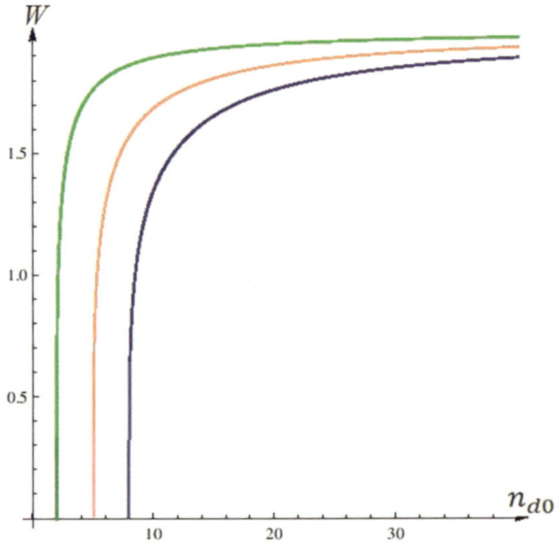

Figure 2.
Variations in DAS width w.r.t. number density of dust grains (n_{do}).

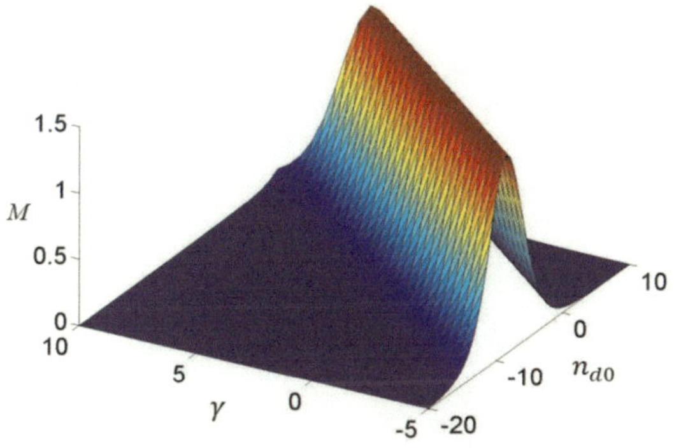

Figure 3.
Variations in phase velocity w.r.t. γ and number density of dust grains (n_{do}).

In this case, the solitary wave will propagate at a higher velocity than in the other cases. From **Figure 3**, it can also be said that the dust acoustic solitary wave will propagate faster with higher amplitude and higher velocity with the increasing value of γ and n_{d0}.

9. Conclusion

In this work, we have presented a brief study on the fundamental properties of nonlinear DA wave structure, the interaction of dust and plasma particles, and the effect of intramolecular attraction and repulsion between these plasma and dust grain particles in inhomogeneous dusty plasma. During this investigation, we have discussed a few fundamental relations of dusty plasma, the effect of dust particles on the nonlinear wave structures, and the basic characteristics of DA waves, which exist in both strongly and weakly coupled dusty plasmas. Here, we have also studied the various aspects of dust-acoustic solitary waves (DASWs) in an inhomogeneous plasma. For studying and analyzing the various aspects of DAWs in inhomogeneous plasmas, the governing fluid equations of plasmas are considered to derive the *Korteweg-de Vries (KdV)* equation. The solution of the KdV equation is obtained as a soliton or solitary wave. The solitary wave solution indicates the various characteristics of DASWs in the inhomogeneous dusty plasma. The above results suggest that the number density of dust particles in DA solitary waves is affected by the negative potentials only. Indeed, we tried to exhibit the effects of dust particles on the nonlinear DA wave structures in inhomogeneous plasma environments. We have also tried to provide a brief idea of dusty plasmas, which is not only fundamental particle of the universe but also dusty plasmas are fundamental concepts like protons and electrons. This chapter also includes a systematic and extensive study on DAWs for inhomogeneous and unmagnetized plasmas.

Author details

Hirak Jyoti Dehingia
Department of Mathematics, Dibrugarh University, Dibrugarh, India

*Address all correspondence to: hirakjyotidehingia11@gmail.com

References

[1] Spitzer L Jr. Review of publications: Physical processes in the interstellar medium. Journal of the Royal Astronomical Society of Canada. 1978;**72**:349

[2] Smith BA, Soderblom L, Batson R, Bridges P, Inge JAY, Masursky H, et al. A new look at the Saturn system: The Voyager 2 images. Science. 1982;**215** (4532):504-537

[3] Spears KG, Robinson TJ, Roth RM. Particle distributions and laser-particle interactions in an RF discharge of silane. IEEE Transactions on Plasma Science. 1986;**14**(2):179-187

[4] Xu W, Song B, Merlino RL, D'Angelo N. A dusty plasma device for producing extended, steady state, magnetized, dusty plasma columns. Review of Scientific Instruments. 1992;**63**(11): 5266-5269

[5] Sheehan DP, Carillo M, Heidbrink W. Device for dispersal of micrometer- and sub micrometer-sized particles in vacuum. Review of Scientific Instruments. 1990;**61**(12):3871-3875

[6] Northrop TG. Dusty plasmas. Physica Scripta. 1992;**45**(5):475

[7] Goertz CK. Dusty plasmas in the solar system. Reviews of Geophysics. 1989;**27** (2):271-292

[8] Rao NN, Shukla PK, Yu MY. Dust-acoustic waves in dusty plasmas. Planetary and Space Science. 1990;**38**(4): 543-546

[9] Shukla PK, Silin VP. Dust ion-acoustic wave. Physica Scripta. 1992;**45** (5):508

[10] Melandso F. Lattice waves in dust plasma crystals. Physics of Plasmas. 1996;**3**(11):3890-3901

[11] Barkan A, D'Angelo N, Merlino RL. Experiments on ion-acoustic waves in dusty plasmas. Planetary and Space Science. 1996;**44**(3):239-242

[12] Barkan A, Merlino RL, D'Angelo N. Laboratory observation of the dust-acoustic wave mode. Physics of Plasmas. 1995;**2**(10):3563-3565

[13] Verheest F. Waves and instabilities in dusty space plasmas. Space Science Reviews. 1996;**77**(3):267-302

[14] Shukla PK. A survey of dusty plasma physics. Physics of Plasmas. 2001;**8**(5): 1791-1803

[15] Verheest F. Waves in dusty space plasmas. Dordrecht: Kluwer Academic Publishers; 2000. p. Xiv + 264

[16] Shukla PK, Mamun AA. Introduction to Dusty Plasma Physics. Bristol: Institute of Physics; 2002

[17] Anowar MGM, Mamun AA. Effects of two-temperature electrons and trapped ions on multidimensional instability of dust-acoustic solitary waves. IEEE Transactions on Plasma Science. 2009;**37**(8):1638-1645

[18] Baluku TK, Hellberg MA. Kinetic theory of dust ion acoustic waves in a kappa-distributed plasma. Physics of Plasmas. 2015;**22**(8):083701

[19] Alinejad H, Khorrami V. Effects of polarized debye sheath and trapped ions on solitary structures in a strongly coupled inhomogeneous dusty plasma. IEEE Transactions on Plasma Science. 2017;**46**(4):755-762

[20] Atteya A, El-Borie MA, Roston GD, El-Helbawy AS. Nonlinear dust acoustic

waves in an inhomogeneous magnetized quantum dusty plasma. Waves in Random and Complex Media. 2021;**0**(0): 1-16

[21] Ma JX, Liu J. Dust-acoustic soliton in a dusty plasma. Physics of Plasmas. 1997; **4**(2):253-255

[22] Mamun AA, Cairns RA, Shukla PK. Solitary potentials in dusty plasmas. Physics of Plasmas. 1996;**3**(2):702-704

[23] Mamun AA. Arbitrary amplitude dust-acoustic solitary structures in three-component dusty plasma. Astrophysics and Space Science. 1999;**268**(4):443-454

[24] Bharuthram R, Shukla PK. Large amplitude ion-acoustic solitons in a dusty plasma. Planetary and Space Science. 1992;**40**(7):973-977

[25] Mamun AA, Shukla PK. Cylindrical and spherical dust ion–acoustic solitary waves. Physics of Plasmas. 2002;**9**(4): 1468-1470

[26] Shukla PK. Dust ion-acoustic shocks and holes. Physics of Plasmas. 2000;**7**(3): 1044-1046

[27] Popel SI, Yu MY, Tsytovich VN. Shock waves in plasmas containing variable-charge impurities. Physics of Plasmas. 1996;**3**(12):4313-4315

[28] Melandsø F, Shukla PK. Theory of dust-acoustic shocks. Planetary and Space Science. 1995;**43**(5):635-648

[29] Shukla PK, Mamun AA. Dust-acoustic shocks in a strongly coupled dusty plasma. IEEE Transactions on Plasma Science. 2001;**29**(2):221-225

[30] Farokhi B, Shukla PK, Tsintsadze NL, Tskhakaya DD. Linear and nonlinear dust lattice waves in plasma crystals. Physics Letters A. 1999;**264**(4):318-323

[31] Rehman HU. Electrostatic dust acoustic solitons in pair-ion-electron plasmas. Chinese Physics Letters. 2012; **29**(6):065201

[32] El-Labany SK, El-Taibany WF, El-Samahy AE, Hafez AM, Atteya A. Higher-order corrections to nonlinear dust-ion-acoustic shock waves in a degenerate dense space plasma. Astrophysics and Space Science. 2014; **354**(2):385-393

[33] Atteya A, Sultana S, Schlickeiser R. Dust-ion-acoustic solitary waves in magnetized plasmas with positive and negative ions: The role of electrons superthermality. Chinese Journal of Physics. 2018;**56**(5):1931-1939

[34] El-Taibany WF, Moslem WM, Wadati M, Shukla PK. On the instability of electrostatic waves in a nonuniform electron–positron magnetoplasma. Physics Letters A. 2008;**372**(22):4067-4075

[35] Akhtar N, El-Tantawy SA, Mahmood S, Wazwaz AM. On the dynamics of dust-acoustic and dust-cyclotron freak waves in a magnetized dusty plasma. Romanian Reports in Physics. 2019;**71**: 403

[36] Ur-Rehman H, Mahmood S, Hussain S. Magneto acoustic solitons in pair-ion fullerene plasma. Waves in Random and Complex Media. 2020;**30**(4):632-642

[37] D'Angelo N. Ion-acoustic waves in dusty plasmas. Planetary and Space Science. 1994;**42**(6):507-511

[38] Nakamura Y, Bailung H, Shukla PK. Observation of ion-acoustic shocks in a dusty plasma. Physical Review Letters. 1999;**83**(8):1602

[39] Luo QZ, D'Angelo N, Merlino RL. Experimental study of shock formation

in a dusty plasma. Physics of Plasmas. 1999;**6**(9):3455-3458

[40] Asgari H, Muniandy SV, Wong CS. Effects of strength of dispersion and dust density on the formation of solitons and shocks in unmagnetized dusty plasma. Physics of Plasmas. 2009;**16**(7):073702

[41] Ghosh S, Sarkar S, Khan M, Gupta MR. Nonlinear properties of small amplitude dust ion acoustic solitary waves. Physics of Plasmas. 2000;7(9): 3594-3599

[42] Losseva TV, Popel SI, Golub' AP, Shukla PK. Evolution of weakly dissipative hybrid dust ion-acoustic solitons in complex plasmas. Physics of Plasmas. 2009;**16**(9):093704

[43] Popel SI, Losseva TV, Golub' AP, Merlino RL, Andreev SN. Dust ion-acoustic shocks in a Q machine device. Contributions to Plasma Physics. 2005; **45**(7):461-475

[44] Kruskal M. Nonlinear wave equations. In: Moser J, editor. Dynamical Systems, Theory and Applications. Lecture Notes in Physics. Vol. 38. Berlin, Heidelberg: Springer; 1975. pp. 310-354. DOI: 10.1007/3-540-07171-7_9

[45] Alinejad H. Dust ion-acoustic solitary and shock waves in a dusty plasma with non-thermal electrons. Astrophysics and Space Science. 2010; **327**(1):131-137

[46] El-Taibany WF, El-Bedwehy NA, El-Shamy EF. Three-dimensional stability of dust-ion acoustic solitary waves in a magnetized multicomponent dusty plasma with negative ions. Physics of Plasmas. 2011;**18**(3):033703

[47] Ferdousi M, Miah MR, Sultana S, Mamun AA. Low-frequency electrostatic shock excitations in a multi-component dusty plasma. Brazilian Journal of Physics. 2015;**45**(2):244-250

[48] Ashraf S, Yasmin S, Asaduzzaman M, Mamun AA. Obliquely propagating nonextensive dust-ion-acoustic solitary waves in a dusty magnetoplasma. Astrophysics and Space Science. 2013; **344**(1):145-151

[49] Bacha M, Tribeche M, Shukla PK. Dust ion-acoustic solitary waves in a dusty plasma with nonextensive electrons. Physical Review E. 2012;**85**(5): 056413

[50] Tribeche M, Zerguini TH. Current-driven dust ion-acoustic instability in a collisional dusty plasma with charge fluctuations. Physics of Plasmas. 2001;**8** (2):394-398

[51] Maitra S, Banerjee G. Dust ion acoustic solitary waves in a collisional dusty plasma with dust grains having Gaussian distribution. Physics of Plasmas. 2014;**21**(11):113707

[52] Janaki MS, Dasgupta B, Gupta MR, Som BK. Solitary magnetosonic waves with Landau damping. Physica Scripta. 1992;**45**(4):368

[53] Ghosh S, Chaudhuri TK, Sarkar S, Khan M, Gupta MR. Collisionless damping of nonlinear dust ion acoustic wave due to dust charge fluctuation. Physical Review E. 2002;**65**(3):037401

[54] Ghosh S, Adak A, Khan M. Dissipative solitons in pair-ion plasmas. Physics of Plasmas. 2014;**21**(1):012303

[55] Chatterjee P, Ali R, Saha A. Analytical Solitary Wave Solution of the Dust Ion Acoustic Waves for the Damped Forced Korteweg–de Vries Equation in Superthermal Plasmas. Zeitschrift für Naturforschung A. 2018; **73**(2):151-159

[56] Pakzad HR. Effect of q-nonextensive distribution of electrons on electron acoustic solitons. Astrophysics and Space Science. 2011;**333**(1):247-255

[57] Muñoz V. A nonextensive statistics approach for Langmuir waves in relativistic plasmas. Nonlinear Processes in Geophysics. 2006;**13**(2):237-241

[58] Taruya A, Sakagami MA. Long-term evolution of stellar self-gravitating systems away from thermal equilibrium: connection with nonextensive statistics. Physical Review Letters. 2003;**90**(18): 181101

[59] Latora V, Rapisarda A, Tsallis C. Non-Gaussian equilibrium in a long-range Hamiltonian system. Physical Review E. 2001;**64**(5):056134

[60] Eslami P, Mottaghizadeh M, Pakzad HR. Head-on collision of ion-acoustic solitary waves in a plasma with a q-nonextensive electron velocity distribution. Physica Scripta. 2011;**84**(1): 015504

[61] Ghosh UN, Ghosh DK, Chatterjee P, Bacha M, Tribeche M. Nonplanar ion-acoustic Gardner solitons in a pair-ion plasma with nonextensive electrons and positrons. Astrophysics and Space Science. 2013;**343**(1):265-272

[62] Rehman AU, Ahmad M, Hamza MY. Effect of non-extensivity parameter q on the damping rate of dust ion acoustic waves in non-extensive dusty plasma. Contributions to Plasma Physics. 2019; **59**(1):54-62

[63] El Ghani O, Driouch I, Chatei H. Effects of non-extensive electrons on the sheath of dusty plasmas with variable dust charge. Contributions to Plasma Physics. 2019;**59**(9):e201900030

[64] Bains AS, Tribeche M, Gill TS. Modulational instability of ion-acoustic waves in a plasma with aq-nonextensive electron velocity distribution. Physics of Plasmas. 2011;**18**(2):022108

[65] Shukla PK. Shielding of a slowly moving test charge in dusty plasmas. Physics of Plasmas. 1994;**1**(5):1362-1363

[66] Washimi H, Taniuti T. Propagation of ion-acoustic solitary waves of small amplitude. Physical Review Letters. 1966;**17**(19):996

Chapter 4

A Summary of the Introduction and Importance of Quantum Plasmas

Jyoti

Abstract

In this chapter, we will discuss the quantum plasmas that have their applications mainly in miniaturized semiconductors, optical fibers, waveguides, nanoplasmonics, and astrophysical systems. Quantum plasmas are the least explored field owing to the astronomical applications of classical plasmas. In this chapter we will discuss how quantum plasmas can be studied and which system of equations will be easier to follow. We will discuss the easiest method possible and more popular way to explore the quantum plasmas for the sake of understanding of a new reader to this subject.

Keywords: quantum plasmas, instabilities, fluid equations, normal mode analysis, plasma physics

1. Introduction

Plasma physics examines a system of charged particles that communicate with one another through the application of electromagnetic forces. At the start of the 20th century, researchers' interest in the physics of gas discharges sparked the beginning of the field of study known as plasma physics. Plasmas have been the subject of a significant amount of research, both experimental and theoretical, ever since the end of World War II. This is primarily because of the potential applications of nuclear fusion for both military (the hydrogen bomb) and civilian purposes (the production of energy through controlled thermonuclear fusion). It should not come as a surprise that plasma physics was created simultaneously by both astrophysicists and geophysicists given the widespread belief that around 90 percent of all matter and energy in the observable universe exists in the form of plasma. Plasmas can be observed in a variety of locations, including the atmosphere of the Sun, the magnetosphere of the earth, as well as the interplanetary and interstellar media.

Plasma physics is generally considered to be an entirely classical subject of study. The last ten years have seen a recent uptick in people's interest in plasma systems, which are places where quantum effects play a key role.

Maxwell-Boltzmann statistics are frequently observed in space plasmas and laboratory plasmas because their densities are low even at temperatures that are rather high. At these temperatures, the de Broglie wavelength is not considerably impacted for electrons and ions.

When compared to the interparticle separation that occurs in plasmas, the thermal de Broglie wavelength of ions and electrons is easily observable in dense plasmas that are kept at low temperatures. Under these conditions, the quantum effect becomes more significant than the classical one. This is because of the significant overlap of matter waves that are connected with particles. Ions, degenerate light particles (such electrons and positrons), and holes are said to be in a dense state in quantum plasmas. Due to the large overlap of matter waves associated with particles in such circumstances, the quantum effect becomes more significant than the classical one. Quantum plasmas are described as having ions, degenerate light particles (such as electrons and positrons), and holes in a dense state. HEMT transistors, resonant tunneling diodes, and metallic nanostructures are examples of quantum semiconductor devices that exhibit plasmas with fermionic properties. Astrophysical systems also exhibit these properties (such as neutron stars, and white dwarfs). Quantum diodes, biophotonics, and ultra-small electrical devices can all be seen to share the same features.

2. Comparison of various methodologies for quantum plasmas

Over the past few decades, there has been an extraordinary amount of focus placed on the electrical properties of nanoscale objects [1–4], particularly when such qualities are triggered into action by electromagnetic radiation. This field of research encompasses a wide range of nano-objects, such as semiconductor quantum dots, carbon nanotubes, metallic films, and nanoparticles; novel substances, such as graphene; and meta-materials, the composition of which can be tailored to display certain electromagnetic properties. Impressive potential applications include efficient high-performance computing (efficient storage and transfer of information), nanoplasmonics (optical filters, waveguides) [5, 6], and even in medical sciences (biomedical diagnostics and sensors) [7, 8]. These applications are all possible thanks to the properties of graphene. The electronic responses of such systems, which are frequently in the nonlinear and out-of-equilibrium zones, can be compared to those of a one-component quantum plasma and can be handled at various degrees of approximation. This comparison is possible because the electronic responses of such systems are frequent in these zones.

In order to account for quantum effects, the currently accepted classical theory for plasma needs to undergo some significant revisions. To properly account for this change, a suitable mathematical formalism is required. Several different research approaches have been established for the study of quantum plasmas. Techniques such as time-dependent density functional theory (TDDFT) and time-dependent Hartree–Fock (TDHF) are examples of such techniques. Although TDDFT and TDHF are wavefunction-based methods, they can be restated in the phase-space language that is common among plasma physicists by utilizing Wigner functions. This will allow the methods to be more easily understood.

Both the wavefunction and the phase space approaches need a significant amount of time and memory to employ, which is especially problematic in systems that contain hundreds of electrons.

Fluid models, which may be generated using the applicable kinetic equations (Wigner for a totally quantum approach or Vlasov for semiclassical modelling), offer the possibility of a cheaper alternative [9]. These models can be constructed by employing velocity moments of the distribution function. The fluid technique should be accurate enough to provide important insights into the underlying physical mechanisms while at the same time remaining affordable in terms of processing cost, despite the fact that some information will inevitably be lost throughout the course of this process. This is because some information cannot be recovered once it has been lost. Similar to their classical analogues, the validity of quantum fluid models is limited to long wavelengths in relation to the interparticle distance [10]. This is because quantum fluids behave differently from classical fluids. They are, nevertheless, capable of handling large excitations, Coulomb exchange (an effect associated to the Pauli exclusion principle), electron-electron correlations, and quantum processes such as tunnelling.

In the fluid approach, the dynamics of the electrons are modelled by a series of hydrodynamic equations (continuity, momentum balance, and energy balance), which incorporate quantum effects through the use of the so-called Bohm potential. When compared to Wigner or TDDFT simulations, one can anticipate a significant reduction in the amount of time required for computing. This is because TDDFT methods are required to solve N minus one Schrodinger-like equations, whereas the phase space approach doubles the number of independent variables (positions and velocities).

In the past, the field of condensed-matter physics made use of hydrodynamic models, primarily for semiconductors [11] and, to a lesser extent, for metal clusters [12, 13]. In more recent times, the concept of the quantum fluid has been expanded to include still another significant quality of the electron, which is referred to as its spin [14]. The resulting fluid equations are much more involved than their spinless counterparts, but they may find useful applications in the emerging field of spintronics [15] in condensed-matter physics and in the study of highly polarised electron beams [16] in plasma physics. Both of these fields are related to the study of condensed matter physics. Further applications of quantum fluid theory include dense plasmas, which are created when solid targets interact with intense laser beams [17], warm dense matter experiments [18], quantum nanoplasmonics [19–22], and compact astrophysical objects, such as white dwarf stars [23].

3. Fluid approach for quantum plasmas

Quantum mechanical effects start playing a significant role when the Wigner-Seitz radius (average interparticle distance) $a = \left(\frac{3}{4\pi n}\right)^{1/3}$ is comparable to or smaller than the thermal de Broglie wavelength $\lambda_B = \frac{\hbar}{mV_T}$, where m is the mass of the quantum particles (e.g., degenerate electrons, degenerate positrons, degenerate holes), $V_T = \left(\frac{k_B T}{m}\right)^{1/2}$ is the thermal speed of the quantum particles, T is the temperature, m is the mass, and k_B is the Boltzmann constant, i.e., when

$$n\lambda_B^3 \geq 1, \tag{1}$$

or, equivalently, when the temperature T is comparable to or lower than the Fermi temperature $T_F = \frac{E_F}{k_B}$, where the Fermi energy is

$$E_F = \frac{\hbar^2}{2m}\left(3\pi^2\right)^{2/3} n^{2/3} \tag{2}$$

The relevant degeneracy parameter for the quantum plasma is

$$\frac{T_F}{T} = \frac{1}{2}\left(3\pi^2\right)^{2/3}\left(n\lambda_B^3\right)^{2/3} \geq 1 \tag{3}$$

For typical metallic densities of free electrons, $n \sim 5 \times 10^{22}$ cm^{-3}, we have $T_F \sim 6 \times 10^4$ K, which should be compared with the usual temperature T.

When the plasma particle temperature approaches T_F , one can show, by using a density matrix formalism [24], that the equilibrium distribution function changes from the Maxwell-Boltzmann $\propto \exp(-E/k_BT)$ to the Fermi-Dirac (FD) distribution function

$$f_{FD} = 2(m/2\pi\hbar)^3 \left[1 + \exp\left(\frac{E-\mu}{k_BT}\right)\right]^{-1} \tag{4}$$

where in the nonrelativistic limit the energy is $E = (m/2)v^2 = (m/2)\left(v_x^2 + v_y^2 + v_z^2\right)$. The chemical potential is denoted byμ . The parameter μ/k_BT is large and negative in the nondegenerate limit, and is large and positive in the completely degenerate limit.

It is useful to define the quantum coupling parameters for electron-electron and ion-ion interactions. The electron-electron Coulomb coupling parameter is defined as the ratio of the electrostatic interaction energy $E_{\text{int}} = e^2/a_e$ between electrons and the electron Fermi energy $E_{Fe} = k_BT_{Fe}$, where e is the magnitude of the electron charge and $a_e = (3/4\pi n_e)^{1/3}$ is the mean interelectron distance. We have

$$T_e = \frac{E_{int}}{E_{Fe}} \approx 0.3\left(\frac{1}{n_e\lambda_{Fe}^3}\right)^{2/3} \approx 0.3\left(\frac{\hbar\omega_{pe}}{k_BT_{Fe}}\right)^2 \tag{5}$$

where $\lambda_{Fe} = V_{Fe}/\omega_{pe}$, is the electron Fermi speed, and $\omega_{pe} = (4\pi^2ne^2/m_e)^{1/2}$ the electron plasma frequency. Furthermore, the ion-ion Coulomb coupling parameter is$\Gamma_i = Z_i^2e^2/a_ik_BT_i$, where Z_i is the ion charge state, $a_i = (3/4\pi n_i)^{1/3}$ is the mean inter-ion distance, and T_i is the ion temperature. Since T_e for metallic plasmas could be larger than unity, it is of interest to enquire the role of interparticle collisions on collective processes in a quantum plasma. It turns that the Pauli blocking reduces the collision rate for most practical purposes [25, 26]. Because of Pauli blocking, only electrons with a shell of thickness k_BT about the Fermi surface suffer collisions. For these electrons, the electron-electron collision frequency is proportional to $k_BT = \hbar$. The average collision frequency among all electrons turns out to be [25]

$$v_{ee} = \frac{k_BT^2}{\hbar T_{Fe}} \tag{6}$$

Typically, $v_{ee}\langle\langle\omega_{pe}$ when $T\langle T_{Fe}$, which is relevant for metallic electrons. On the other hand, the typical time scale for electron-ion (lattice) collisions is $\tau_{ei} \cong 10fs$, which is 1 order of magnitude greater than the electron plasma period. Accordingly, a collisionless quantum plasma regime is relevant for phenomena appearing on the time scale of the order of a femtosecond in a metallic plasma.

4. Various systems of equations

4.1 The QHD equation

The nonrelativistic QHD equations [27] have been developed in condensed matter physics [28] and in plasma physics [9, 25]. The nonrelativistic QHD equations are composed of the electron continuity equation

$$\frac{\partial n_e}{\partial t} + \nabla.(u_e n_e) = 0 \tag{7}$$

the electron momentum equation

$$me\left(\frac{\partial u_e}{\partial t} + u_p.\nabla u_e\right) = e\nabla\varphi - \frac{1}{n_e}\nabla p_e + F_Q = 0 \tag{8}$$

and Poisson's equation

$$\nabla^2\varphi = 4\pi e(n_e - Z_i n_i) \tag{9}$$

In a quantum plasma with nonrelativistic degenerate electrons, the quantum statistical pressure in the zero electron temperature limit can be modeled as [9, 29]

$$p_e = \frac{m_e V_{Fe}^2 n_0}{3}\left(\frac{n_e}{n_0}\right)^{(D+2)/D} \tag{10}$$

where D is the number of space dimension of the system and $V_{Fe} = (\hbar/m_e)(3\pi^2 n_e)^{1/3}$ is the electron Fermi speed.

4.2 The NLS-Poisson equation

For investigating the nonlinear properties of dense quantum plasmas, it is also possible to work with a NLS equation. Hence, by introducing the wave function

$$\psi(r, t) = \sqrt{n_e(r, t)}\exp\left(\imath\frac{S_e(r,\ t)}{\hbar}\right) \tag{11}$$

where S_e is defined according to $m_e u_e = \nabla S_e$ and $n_e = |\psi|^2$, it can be shown that continuity equation and equation for momentum can be cast into a NLS equation [9, 25]

$$\imath\hbar\frac{\partial\psi}{\partial t} + \frac{\hbar^2}{2m_e}\nabla^2\psi + e\varphi\psi - \frac{meV_{Fe}^2}{2n_0^2}|\psi|^{4/D}\psi = 0 \tag{12}$$

where the electrostatic field φ arising from the charge distribution of N electrons is determined from Poisson's equation

$$\nabla^2\varphi = 4\pi e\left(|\psi|^2 - Z_i n_i\right) \tag{13}$$

We note that the third and fourth terms on the left-hand side of Eq. (12) represent the nonlinearities associated with the nonlinear coupling between the electrostatic potential and the electron wave function and the nonlinear quantum statistical pressure, respectively.

5. Numerical mode analysis method for investigation of two stream instabilities

Let us study the basic fluid equation for quantum plasmas to investigate the Two Stream instabilities using Normal Mode Analysis Method (NMA) which is one of the basic and simplest way to understand the instabilities in plasmas. For the sake of understanding and simplification let us rewrite the Fluid equations again, which includes continuity equations and momentum equations for a plasma model which is unmagnetized and possess non degenerate ions and moves with non-relativistic speed.

5.1 Basic equations

In an unmagnetized, non-relativistic collision-less dense plasma, the QHD equation for a degenerate electron under the effect of an electrostatics potential due to non-degenerate ions [30, 31] that move at a non-relativistic speed is self-possessed of a continuity equation and a momentum equation for a degenerate electron. This is the case because non-degenerate ions move at a non-relativistic electrons

$$\frac{\partial n_e}{\partial t} + \vec{\nabla}\cdot\left(n_e\vec{u}_e\right) = 0 \tag{14}$$

$$m_e n_e\left[\frac{\partial \overrightarrow{u_e}}{\partial t} + \left(\vec{u}_e\cdot\vec{\nabla}\right)\vec{u}_e\right] - en_e\vec{\nabla}\varphi + \vec{\nabla}P_e - n_e\overrightarrow{F_B} = 0 \tag{15}$$

Where

$\vec{F}_B = F_{Bx}\wedge i + F_{Bz}\wedge k = \frac{\hbar^2\vec{\nabla}}{2m_e}\left(\frac{\nabla^2\sqrt{n_e}}{\sqrt{n_e}}\right)$ is the force arising due to Bohm potential, which is self – consistent potential depends upon the amplitude of wave function of matter wave associated with degenerate electrons.

Along with Poisson equation

$$\varepsilon_0\nabla^2\varphi - \frac{e}{\varepsilon_0}(n_e - Z_i n_i) = 0 \tag{16}$$

In unmagnetized quantum plasma, the dynamics of non-degenerate ions, is described by the QHD equation in non-relativistic region, which includes continuity equation and momentum equation for non-degenerate ions

$$\frac{\partial n_i}{\partial t} + \vec{\nabla}\cdot\left(n_i\vec{u}_i\right) = 0 \tag{17}$$

$$\left(1+\tau_m\frac{\partial}{\partial t}\right)\left[\left(\frac{\partial}{\partial t}+\vec{u}.\vec{\nabla}\right)\vec{u} + \frac{Z_i e}{m_i}\vec{\nabla}\varphi + \frac{\gamma_i k_B T_i}{m_i n_i}\vec{\nabla}n_i\right] - \frac{\eta}{\rho_i}\nabla^2 u_i - \frac{\left(\xi+\frac{\eta}{3}\right)}{\rho_i}\vec{\nabla}\left(\vec{\nabla}.\overrightarrow{u_i}\right) = 0 \tag{18}$$

Where Z_i is atomic number of ions, φ is electrostatics potential due to space charge distribution of electron-ion system., n_e,n_i are number density of electrons and ions m_e,m_i are Mass of electrons and ions, u_e, u_i are velocity of electrons and ions, ρ_i is mass density of ion, γ_i is Adiabetic index of ion, τ_m relaxation time of ion η and ξ = 0.1 and 2.7 = Bulk ion visco-elastic coefficient for ions while the rest of physical quantities have their usual meaning.

Here momentum equation includes the quantum statistical pressure term for degenerate electron [32] which behaves like a Fermi gas at zero-electron temperature, which depends upon Fermi speed of degenerate Fermi electron gas and degree of space-dimension of electron [33, 34].

At zero-electron temperature, quantum statistical pressure is

$$P_e = \frac{m_e (V_{Fe})^2 n_o}{3} \left(\frac{n_e}{n_o}\right)^{\left(\frac{2}{f}+1\right)} \tag{19}$$

Where

$V_{Fe} = \frac{\left(3\pi^2 \hbar^3 n_e\right)^{\frac{1}{3}}}{m_e}$ = Fermi speed of degenerate electron
f = degree of space-dimension of degenerate electron.
n_o = number density of electron in equilibrium state.

5.2 Mathematical analysis

The continuity and momentum equations for electrons and ions, as well as Poisson's equation, are solved using Normal Mode Analysis and linearization, where the physical quantities (density, velocity, and electrostatic potential) n_i ,n_e, u_{ix}, u_{iy}, u_{iz}, u_{ex}, u_{iy}, u_{iz} and φ are expressed as $P = P_0 + P_1$ where P_0 and P_1 represent the unperturbed and perturbed part respectively. Due to their tiny size, higher order perturbed terms, particularly those of the second order, are disregarded during linearization. After calculating the perturbed velocities for ions and electrons, u_{i1} and u_{e1}, from momentum equations, these values are substituted into the expression for perturbed densities of ions and electrons obtained from continuity equations under the assumption that the unperturbed densities of ions and electrons are constant with respect to the electron oscillation period, such that-

$$\frac{\partial n_{i0}}{\partial t} = \frac{\partial n_{e0}}{\partial t} = 0 \tag{20}$$

Also unperturbed components of velocities of ions and electrons are considered constant in space-time, so that

$$\frac{\partial u_{ix0}, u_{iy0}, u_{iz0}}{\partial x, z} = \frac{\partial u_{ex0}, u_{ey0}, u_{ez0}}{\partial x, z} = 0 \tag{21}$$

$$\frac{\partial u_{ix0}, u_{iy0}, u_{iz0}}{\partial t} = \frac{\partial u_{ex0}, u_{ey0}, u_{ez0}}{\partial t} = 0 \tag{22}$$

After applying all this conditions, the density of electrons and ions are evaluated in term of perturbed potential φ_1 and Using these values of np1 and ne1 in Poisson's Eq. (18), we get the following equation in φ_1

$$\varphi_1 = \frac{\varphi_N}{\varphi_D} = \frac{\varphi_{NR} + i\varphi_{NI}}{\varphi_{DR} + i\varphi_{DI}} \tag{23}$$

Where

$$\varphi_N = \varphi_{NR} + i\varphi_{NI} \tag{24}$$

and

$$\varphi_D = \varphi_{DR} + i\varphi_{DI} \tag{25}$$

Real and imaginary parts of numerator and denominator are written as:

$$\varphi_{NR}= \begin{bmatrix} \omega^8 G_{\mathrm{NR1}} + \omega^7 G_{\mathrm{NR2}} + \omega^6 G_{\mathrm{NR3}} + \omega^5 G_{\mathrm{NR4}} + \omega^4 G_{\mathrm{NR5}} \\ +\omega^3 G_{\mathrm{NR6}} + \omega^2 G_{\mathrm{NR7}} + \omega^1 G_{\mathrm{NR8}} + \omega^0 G_{\mathrm{NR9}} \end{bmatrix} \tag{26}$$

$$\varphi_{NI} = \begin{bmatrix} \omega^8 G_{\mathrm{NI1}} + \omega^7 G_{\mathrm{NI2}} + \omega^6 G_{\mathrm{NI3}} + \omega^5 G_{\mathrm{NI4}} + \omega^4 G_{\mathrm{NI5}} \\ +\omega^3 G_{\mathrm{NI6}} + \omega^2 G_{\mathrm{NI7}} + \omega^1 G_{\mathrm{NI8}} + \omega^0 G_{\mathrm{NI9}} \end{bmatrix} \tag{27}$$

$$\varphi_{DR} = \begin{bmatrix} \omega^{10} G_{\mathrm{DR1}} + \omega^9 G_{\mathrm{DR2}} + \omega^8 F G_{\mathrm{DR3}} + \omega^7 G_{\mathrm{DR4}} \\ +\omega^6 G_{\mathrm{DR4}} + \omega^5 G_{\mathrm{DR5}} + \omega^4 G_{\mathrm{DR6}} + \omega^3 G_{\mathrm{DR7}} \\ +\omega^2 G_{\mathrm{DR8}} + \omega^1 G_{\mathrm{DR9}} + \omega^0 G_{\mathrm{DR10}} \end{bmatrix} \tag{28}$$

$$\varphi_{DI} = \begin{bmatrix} \omega^{10} G_{\mathrm{DI1}} + \omega^9 G_{\mathrm{DI2}} + \omega^8 G_{\mathrm{DI3}} + \omega^7 G_{\mathrm{DI4}} \\ +\omega^6 G_{\mathrm{DI3}} + \omega^5 G_{\mathrm{DI4}} + \omega^4 G_{\mathrm{DI5}} + \omega^3 G_{\mathrm{DI6}} \\ +\omega^2 G_{\mathrm{DI7}} + \omega^1 G_{\mathrm{DI8}} + \omega^0 G_{\mathrm{DI9}} \end{bmatrix} \tag{29}$$

Here, G's are coefficients of different powers of ω, which are functions of unperturbed quantities.

Eq. (29) relates the perturbed quantity ϕ_1 with the unperturbed quantities. Since the first-order quantity cannot be explicitly expressed in terms of zeroth-order quantities [35], we make the RHS of this equation indeterminate by putting numerator and denominator to be individually zero. Thus, we get two equations of order seven for frequency of oscillations. These equations are solved numerically for the values of growth rate (imaginary part of ω) by giving typical values of n_{e0}, n_{p0}, u_{x0}, u_{y0}, u_{z0}, v_{x0}, v_{y0}, v_{z0}, and T_e. The real roots of these equations give rise to the propagating waves, which are not of our concern here. The positive imaginary part of the complex root gives the instability, whereas its negative part gives the damped wave. It means the oscillations are unstable when $\mathrm{Im}(\omega) < 0$.

Here **Figure 1**, presents the variation of normalized growth rate of the instabilities occurring into the plasma. From above mentioned polynomial equation in terms of ω, several equations will have complex conjugate roots and same may not have. The growth rate can be analyzed with variation in other typical parameters like ion density, angle of propagation in case of magnetized plasmas, with ion temperature in case of warm plasmas and also with respect to variation in magnetic field in magnetized plasmas.

6. Future perspectives

In the past many derived the dielectric constant for the high-frequency (in comparison with the ion plasma frequency) ES waves [36, 37] and the refractive index for EM waves [38] by using a quantum kinetic theory based on the Wigner and Poisson-Maxwell equations in an unmagnetized quantum plasmas. But no one has found the same in magnetized plasma. In 2D electrons, at very high magnetic field, electron density is quantized which is associated with quantized Hall resistance. After solving this problem we will be able to understand the features of quantum oscillations of electrons and possible formation of bound states of electrons in the presence an external magnetic field.

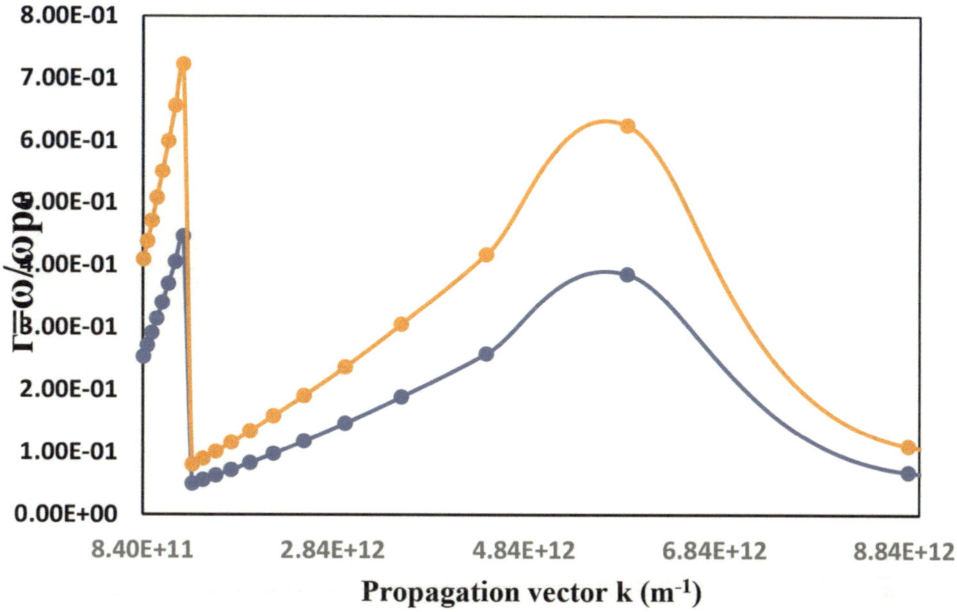

Figure 1.
Variation of growth rate Γ with Propagation Vector (k in m^{-1}) When $\theta = 45°$, $\varphi_o = 0.05$, $n_{eo} = 5 \times 1017\ m^{-3}$, $m_e = 9.1 \times 10^{-31}$kg, $m_i = 1.67 \times 10^{-27}$ kg, $U_{ixo} = U_{iyo} = U_{izo} = 0.25Cs$, $U_{exo} = U_{eyo} = U_{ezo} = 5\ Cs$, $T_e = 0.25eV$, $T_i = 0$ and $Z_i = 1$.

In the past researchers have given dielectric constant without quantum statistical pressure to investigate the screening and wake potentials around a test charge in an electron ion quantum plasma. The wake potential behind a test charge arises due to collective interactions between a test charge and the ion oscillation. The screening potential can be used to study the doubly excited resonance states of He and H atoms embedded in a quantum plasma and lattice waves in 2D hexagonal quantum plasma crystals. In this problem we will deduce potentials around a moving test charge in a quantum plasma with inclusion of statistical pressure term.

Author details

Jyoti
Government College for Women, Rewari, Haryana, India

*Address all correspondence to: jyoti.malik@alumni.iitd.ac.in

References

[1] Voisin C, Christoflos D, Del Fatti N, Vallée F, Prével B, Cottancin E, et al. Size-dependent electron-electron interactions in metal nanoparticles. Physical Review Letters. 2000;**85**(10):2200-2203

[2] Bigot JY, Halté V, Merle JC, Daunois A. Electron dynamics in metallic nanoparticles. Chemical Physics. 2000;**251**(1):181-203

[3] Manfredi G, Hervieux PA, Tanjia F. Quantum hydrodynamics for nanoplasmonics. In: Plasmonics: Design, Materials, Fabrication, Characterization, and Applications XVI. Bellingham: International Society for Optics and Photonics; 2018. p. 107220B

[4] Maniyara RA, Rodrigo D, Yu R, Canet-Ferrer J, Ghosh DS, Yongsunthon R, et al. Tunable plasmons in ultrathin metal flms. Nature Photonics. 2019;**13**(5):328-333

[5] Stockman MI. Nanoplasmonics: The physics behind the applications. Physics Today. 2011;**64**(2):39-44

[6] Moreau A, Ciracì C, Mock JJ, Hill RT, Wang Q, Wiley BJ, et al. Controlledrefectance surfaces with flm-coupled colloidal nanoantennas. Nature. 2012;**492**(7427):86-89

[7] Hainfeld JF, Slatkin DN, Smilowitz HM. The use of gold nanoparticles to enhance radiotherapy in mice. Physics Medical Biology. 2004; **49**(4904):309-315

[8] Tatsuro E, Kagan K, Naoki N, Ha Minh H, Do-Kyun K, Yuji Y, et al. Multiple LabelFree detection of antigen antibody reaction using localized surface plasmon resonance based core shell structured nanoparticle layer nanochip. Analytical Chemistry. 2006;**78**:6465-6475

[9] Manfredi G, Haas F. Self-consistent fuid model for a quantum electron gas. Physical Review B. 2001;**64**(7):075316

[10] Khan SA, Bonitz M. Quantum Hydrodynamics: Complex Plasmas. New York: Springer; 2014. pp. 103-152

[11] Müller T, Parz W, Strasser G, Unterrainer K. Infuence of carrier-carrier interaction on time-dependent intersubband absorption in a semiconductor quantum well. Physical Review B. 2004;**70**(15):155324

[12] Domps A, Reinhard PG, Suraud E. Theoretical estimation of the importance of two-electron collisions for relaxation in metal clusters. Physical Review Letters. 1998;**81**(25):5524-5527

[13] Banerjee A, Harbola MK. Hydrodynamic approach to time-dependent density functional theory; response properties of metal clusters. The Journal of Chemical Physics. 2000;**113**:5614

[14] Brodin G, Marklund M. Spin magnetohydrodynamics. New Journal of Physics. 2007;**9**(8):277

[15] Hirohata A, Yamada K, Nakatani Y, Prejbeanu IL, Diény B, Pirro P, et al. Review on spintronics: principles and device applications. Journal of Magnetism and Magnetic Materials. 2020;**509**:166711

[16] Wu Y, Ji L, Geng X, Yu Q, Wang N, Feng B, et al. Polarized electron-beam acceleration driven by vortex laser pulse. New Journal of Physics. 2019;**11**:073052

[17] Kremp D, Bornath T, Hilse P, Haberland H, Schlanges M, Bonitz M. Quantum kinetic theory of laser plasmas. Contributions to Plasma Physics. 2001; **41**(2–3):259-262

[18] Dornheim T, Groth S, Bonitz M. The uniform electron gas at warm dense matter conditions. Physics Reports. 2018;**744**:1-86

[19] Ciracì C, Pendry JB, Smith DR. Hydrodynamic model for plasmonics: a macroscopic approach to a microscopic problem. Chemical Physics. 2013;**14**(6): 1109-1116

[20] Ciracì C, Della Sala F. Quantum hydrodynamic theory for plasmonics: impact of the electron density tail. Physical Review B. 2016;**93**:205405

[21] Baghramyan HM, Della Sala F, Ciracì C. Laplacian-level quantum hydrodynamic theory for plasmonics. Physics Review. 2021;**X 11**:011049

[22] Toscano G, Straubel J, Kwiatkowski A, Rockstuhl C, Evers F, Xu H, et al. Resonance shifts and spill-out efects in self-consistent hydrodynamic nanoplasmonics. Nature Communications. 2015;**6**(1):1-11

[23] Uzdensky DA, Rightley S. Plasma physics of extreme astrophysical environments. Reports on Progress in Physics. 2014;**77**(3):036902

[24] Bransden BH, Joachain CJ. Quantum Mechanics. 2nd ed. Essex: Pearcon Education Ltd.; 2000

[25] Manfredi G. How to model quantum plasmas. *arXiv*. 2005. DOI: 10.48550/ arXiv.quant-ph/0505004

[26] Son S, Fisch NJ. Current-drive efficiency in a degenerate plasma. Physical Review Letters. 2005;**95**:225002

[27] Wilhelm R. The Garching Belt Pinch experiments. Nuclear Fusion. 1985;**25**:1055

[28] Carl L. Gardner and Christian Ringhofer, Smooth quantum potential for the hydrodynamic model. Physical Review E. 1996;**53**:157

[29] Crouseilles N, Hervieux P-A, Manfredi G. Quantum hydrodynamic model for the nonlinear electron dynamics in thin metal films. Physical Review B. 2008;**78**:15541

[30] Shukla P, Eliasson B. Collo-quium: Nonlinear collective interactions in quantum plasmas with degenerate electron fluids. Reviews of Modern Physics. 2011;**83**(3):885

[31] Shukla PK, Eliasson B. Nonlinear aspects of quantum plasma physics. Physics Uspekhi. 2010;**53**(1):51-76

[32] Shukla PK, Eliasson B. Novel attractive force between ions in quantum plasmas. Physical Review Letters. 2012; **108**(16):165007

[33] Rehman S. Linear and nonlinear quantum ion acoustic waves in a plasma with positive, negative ions and Fermi electron gas. Physics of Plasmas. 2010; **17**(6):062303

[34] Singh SS, Misra KP. Two stream instabilities in unmagnetized nonrelativistic quantum plasma. Plasma Research Express. 2022;**4**(2):025006

[35] Moldabekov ZA et al. Dynamical structure factor of strongly coupled ions in a dense quantum plasma. Physical Review E. 2019;**99**(5):053203

[36] Klimontovich YL, Silin V. To the theory of excitation spectra of macroscopic systems. Doklady Akademii Nauk SSSR. 1952;**82**:361

[37] Bohm D, Pines DJPR. A collective description of electron interactions: III. Coulomb interactions in a degenerate electron gas. 1953;**92**(3):609

[38] Burt P, Wahlquist HDJPR. Spin and exchange corrections to plasma dispersion relations. 1962;**125**(6):1785

Section 3

Particle Acceleration in the Venus Wake

Chapter 5

Particle Acceleration in the Venus Plasma Wake

Hector Pérez-de-Tejada and R. Lundin

Abstract

Measurements conducted with the PVO and the VEX spacecraft have shown that the solar wind that streams around the Venus ionosphere produces plasma vortex-shaped structures across its wake. Those features are a varying property whose width gradually decreases with the downstream distance along the wake. Further studies suggest that as the width of the vortex structures becomes smaller with distance and since the total energy of the particles involved in the vortex motion should be maintained, the energy of particles that move in the central part of the wake should be enhanced. As a result, there should be a continuous energization of the planetary ions that are carried off by the solar wind from the Venus ionosphere; namely, accelerated planetary ions should be measured along the Venus wake. Different from measurements conducted in the Venus near wake where planetary O+ ions move with speeds smaller than those of the solar wind, the conditions far downstream along the wake imply that as a result of the gradual decrease of the vortex width with distance downstream from Venus, the planetary ions that stream in that direction acquire larger speed values and thus become accelerated.

Keywords: venus plasma wake, particle acceleration, pioneer venus orbiter (PVO), venus express (VEX), vortex structures

1. Introduction

Measurements conducted with the Pioneer Venus Orbiter (PVO) and the Venus Express (VEX) spacecraft in orbit around Venus have provided evidence for the presence of vortex structures in the Venus wake and that reveal process similar to those that occur in similar fluid dynamic problems. An initial indication of the presence of vortex motion in the Venus wake was provided by changes in the direction of the velocity vectors of ions reported from the plasma measurements of the PVO [1, 2], and that was better described from the velocity vectors of the H+ and O+ ions obtained with the VEX data projected on a plane transverse to the wake direction reproduced in **Figure 1** [3]. Measurements made for both ion populations describe a counterclockwise rotation as seen from the wake and that extends across a large part of Venus cross section. The velocity distribution is oriented on a geometry different from what would be produced by magnetic tension forces along the field lines from the Venus polar regions. Instead, flow motion is dominant to produce the vortex shape in the particle displacement and that derives from effects related to the rotation of

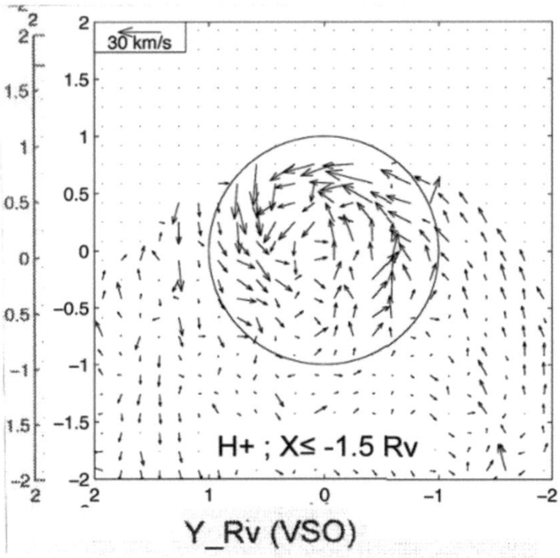

Figure 1.
(left panel) Velocity vectors of the H+ ions projected on the YZ-plane compiled from the VEX measurements (≈ 1–300 eV). Notice the formation of a tail ward moving vortex (helix) in the near wake at $X \leq -1.5\ R_V$. (right panel) Velocity vectors of the O+ ions projected on the YZ-plane compiled from the VEX measurements. Notice the formation of a tail ward moving vortex (helix) in the near wake at $X \leq -1.5\ R_V$ (from [3]).

the Venus ionosphere. Further studies also showed a downward displacement of the trans-terminator flow in the Venus upper ionosphere [4, 5] and on the position of ionospheric holes [6] that were later interpreted as resulting from plasma channels that extend downstream from the Venus polar regions [7].

Particle acceleration of planetary O+ ions in the Venus wake has been a source of information regarding the fact that their measured speed values (20–30 km/s) are significantly smaller than the speed of the solar wind ions (200–300 km/s) in the near wake. Such smaller values are not expected from the acceleration of the planetary ions produced through the solar wind convective electric field where both plasma populations should rapidly acquire similar speeds. Instead, proton cyclotron waves produced through wave particle interactions [8] may lead the O+ ions to acquire larger speed values as they move along the Venus tail, thus becoming gradually accelerated. Information will be presented to stress that magnetic field forces are not sufficient to produce that response and thus that other mechanisms are required to produce the measured well-order displacement of the accelerated planetary ions. As a whole it will be suggested that through wave-particle interactions, the dynamics of vortex structures whose width decreases with distance downstream from Venus will produce a continuous enhancement of the particle speed for fluxes that remain moving along the wake direction.

2. Vortex data from the VEX measurements

Data from 8 VEX orbits obtained during 2006 and 2009 as the spacecraft moved into and left a vortex structure are summarized in **Table 1** and that were selected to produce the segments placed in **Figure 2**. As a whole there is a general difference in the position of the segments, which occurred closer to Venus (which is situated at the right side) in those corresponding to the 2009 orbits (thick underline) and that

DOI: http://dx.doi.org/10.5772/intechopen.109039

Date	Inbound												
	UT	X	Y	Z	v	UT	X	Y	Z	v	δ	Δ	ΔT
Aug 22/06	01.45	−2.83	−0.15	−0.83	25	01:55	−2.58	−0.15	−0.40	25	0.25	0.43	10
Aug 23/06	01:57	−2.75	−0.07	−0.60	15	02:05	−2.40	−0.07	−0.28	16	0.35	0.32	8
Aug 24/06	02.10	−2.45	−0.01	−0.26	30	02:18	−2.03	−0.01	0.20	30	0.42	0.46	8
Aug 28/06	02:22	−2.49	−0.28	−0.38	20	02:28	−2.20	0.24	0.07	20	0.29	0.31	6
Sep 19/09	01:54	−2.51	−0.04	−1.04	15	02:03	−2.20	−0.05	−0.55	15	0.31	0.49	9
Sep 21/09	02:03	−2.30	0.08	−0.65	15	02:13	−1.95	0.06	−0.12	15	0.35	0.53	10
Sep 25/09	02:15	−2.15	0.33	−0.45	20	02:27	−1.60	0.23	0.21	28	0.55	0.66	12
Sep 26/09	02:05	−2.47	0.46	−1.06	20	02:20	−1.83	0.32	−0.01	20	0.64	1.05	15

Table 1.
Corrected VEX coordinates (in R_V) together with the speed v of the planetary O+ ions (in km/s) measured during the inbound (left columns) and the outbound (right columns) crossings of vortex structures in four orbits of 2006 and four orbits of 2009. The columns in the far right indicate the width of the vortex structure in the X and in the Z directions and ΔT the time difference between both crossings.

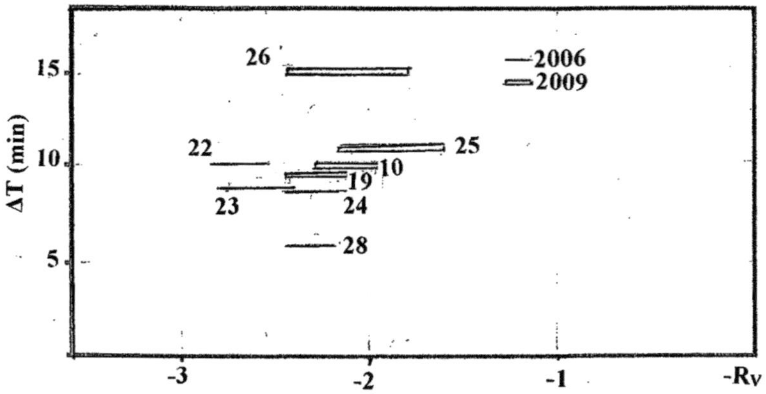

Figure 2.
*Segments measured between the inbound and the outbound crossings of vortex structures by VEX in the eight orbits included in **Table 1** (they are identified by changes in the particle flux intensity measured in the energy spectra of the O+ ions). Their corrected position along the X-axis shows that the 2009 orbits (marked with a double underline) are located closer to Venus, and also that the width value in the 2006 orbits (marked with a single underline) are mostly smaller since they have lower ΔT values and are encountered further downstream along the wake The numbers by the side of each segment state the day of their crossing date in **Table 1** (improved from [9]).*

represent conditions suitable to a minimum in the solar cycle. The segments derived from the 2006 orbits (single underline) are located further downstream along the wake [9]. Also, there is an indication that the time width ΔT between the inbound and the outbound crossings marked at the vertical coordinate in **Figure 2** is smaller in the 2006 orbits which trace the wake farther away from Venus (that transition is derived from the δ and the Δ values annotated in **Table 1** together width their time width ΔT obtained from their measurements). The numbers at the side of each segment indicate the day in the date of the orbits in **Table 1**.

A similar conclusion can also be reached by plotting the inbound and the outbound crossings of vortex structures in those orbits and that is projected on the XZ plane as shown in **Figure 3**. The position of the data points is displaced to lower −Z values with increasing distance downstream from Venus. The two sets of data with four orbits each indicate a different displacement of the vortex structure along the −Z direction. In all the data points, there is a similar displacement in the position of the vortex structure toward the southern hemisphere with increasing distance downstream from Venus. However, there is a general preference in those features to occur closer to Venus in the 2009 measurements since the VEX passage across the Z = 0 axis is by X = −1.8 R_V in that set while it reaches X = − 2.2 R_V in the 2006 measurements. This difference implies that the vortex structures are located closer to Venus during solar cycle minimum conditions by 2009, and that their position along the wake varies along that cycle. The persistent location of the vortex crossings following a similar trend for both the inbound and the outbound passes, which is suitable for 2006 and 2009, supports the view that it is a common phenomenon. It is also to be noted that the close crossing position in orbits labeled 22, 23, 24, and 28 in **Figure 3** (which correspond to the day of the date where they were made during 2006) gives the rate of change of the vortex size with distance in the X and the Z coordinates. In fact, since ΔZ ~ 0.2 R_V and ΔX ~ 0.3 R_V across those orbits, it is possible to suggest that the vortex width changed at a rate with a significant (~ 30°) angle in that part of the wake.

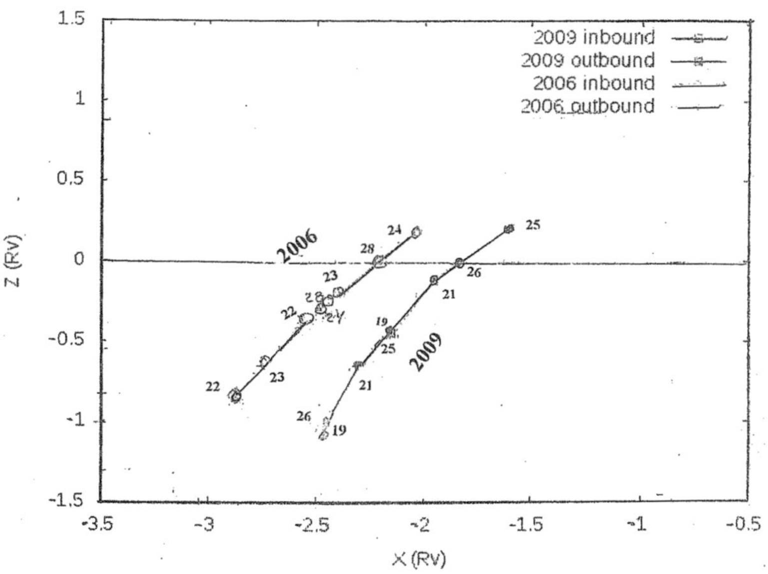

Figure 3.
Corrected position of the VEX spacecraft projected on the XZ plane during its entry (inbound) and exit (outbound) through a corkscrew plasma structure in several orbits. The two traces correspond to the four orbits in 2006 and the four orbits in 2009 listed in ***Table 1****. The numbers at the site of each crossing state the day of their crossing date in that* ***Table 1****.*

An extended study of those results was conducted by considering VEX orbits that exhibit an indication of vortex structures throughout the operation time of that spacecraft (between 2006 and 2013). Even though the plasma conditions differed significantly among them, it was possible to collect a number of orbits for each year of operation and that led to their general distribution reproduced in **Figure 4** [10]. In particular, they describe variations in their width as follows: For each orbit there is a segment bounded by the entry and exit of the spacecraft through a vortex with a number that marks the two last digits of the year when measurements were made (they include the four orbits for 2006 and for 2009 that were discussed in **Figures 2** and **3**). The length of that segment corresponds to the time width of the vortex structure and is marked by its position placed on the vertical coordinate; that is, segments positioned at high values in that axis indicate wide vortices while those situated at low values correspond to thin vortices. Most notable is that the segments identify two different regions; one corresponding to orbits before the minimum solar cycle conditions (between 2006 and 2009) and the other to orbits that occurred during and after those conditions (between 2009 and 2013). Two big circles select schematically different sets of orbits that are located either far away from Venus between 2006 and 2009 (left circle) and those that are placed closer to Venus (between 2009 and 2013 (right circle) during solar cycle minimum conditions. The implication here is that as in **Figures 2** and **3,** the vortex structures occur closer to Venus during minimum solar cycle conditions.

Equally important is that the time width ΔT (segment length) in the 2006: 2009 orbit range is clearly smaller (placed at lower values along the vertical coordinate within the left circle) than that in the 2009: 2913 orbit range (larger values in the right circle). In agreement with both traces in **Figure 3** the vortex structures identified between 2006 and 2009 (left circle in **Figure 4**) occur farther away from Venus than those detected between 2009 and 2013 (right circle). As a result, the thickness

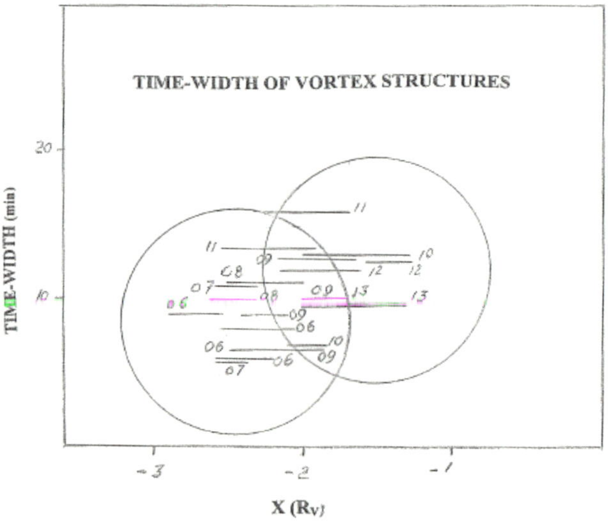

Figure 4.
Corrected values of the time-width (in minutes) between the VEX inbound and outbound crossings of vortex structures as a function of the X-distance (R_V) downstream from Venus that were measured in 20 orbits. The numbers at the side of each segment represent the two last digits of the year when measurements were made in different orbits between 2006 and 2013 (four orbits were examined during 2006 and also during 2009). The two circles confine orbits between 2006 and 2009 (left circle) prior to a solar cycle minimum and those between 2009 and 2013 (right circle) during and after that period [9].

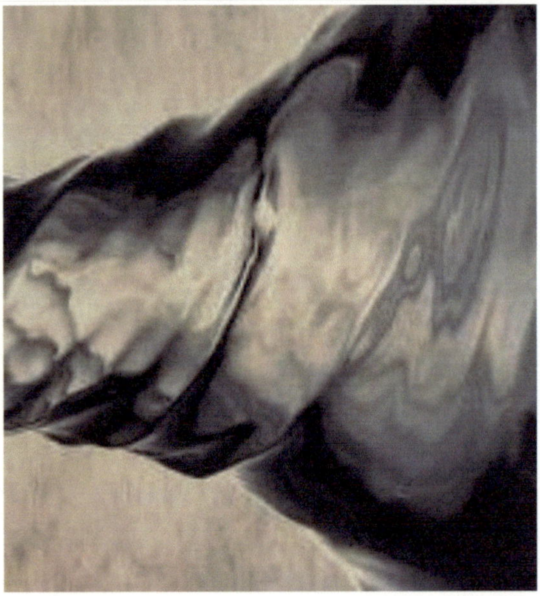

Figure 5.
View of a corkscrew vortex flow in fluid dynamics. Its geometry is equivalent to that of a vortex flow in the Venus wake with its width and position varying during the solar cycle. Near minimum solar cycle conditions the vortex is located closer to Venus (right side) and there are indications that its width becomes smaller with increasing distance downstream from the planet. Such is the case for the 2006 orbits (marked with a single underline in ***Figure 2****) and that were conducted before the solar cycle minimum at 2009–2010 thus implying that the vortex flow becomes thinner when it is detected further downstream along the wake [9].*

of the vortex structures located far away from Venus (left circle) becomes smaller with increasing distance along the wake, thus implying a corkscrew shape similar to that represented schematically in **Figure 5** for a flow configuration behind an object immersed in a fluid and where it thins out with the downstream distance [11].

3. Particle acceleration in the Venus wake

The tendency of the width of vortex structures to decrease with distance downstream from Venus indicated in **Figure 2** has implications regarding its effects on the motion of the planetary ions that stream within them. Most notable is that their kinetic energy around such features depends on the scale size of the vortices and if the latter become smaller with the downstream distance as produced by the expansion of the solar wind into the wake, the energy released should be assimilated by particles that remain moving in the smaller-size vortices. In addition, such particles have directed motion along the wake, and as a result, some of that energy should contribute to enhance their speed in that direction. It is thus possible that a fraction of planetary ions dragged along by the solar wind may be gradually accelerated along the Venus wake.

A velocity variation in that sense can be inferred from the speed profile of the planetary ions shown in both panels of **Figure 6** [12] where in addition to a gradual decrease of their speed with altitude by and above the ionopause (by ~10^3 km), there is a sudden and unexpected change in the altitude gradient by ~5 10^3 km and that produces enhanced speed values from ~10 km/s by ~5 10^3 km to ~40 km/s by ~10^4 km and that peak at that altitude. That variation is nearly the same in both panels of **Figure 6** and suggests the participation of a different process. In particular, larger speed values imply the acceleration of the planetary ions that can be estimated in terms of a change in the cross section of their motion. The acceleration of those particles can be estimated in terms of the cross section of the vortex structure encountered along the spacecraft trajectory. In fact, since the width of a vortex structure decreases with distance along the wake as a result of the expansion of the solar wind into that region (Pérez-de-Tejada, 1986), the total energy content of the vortex motion should be

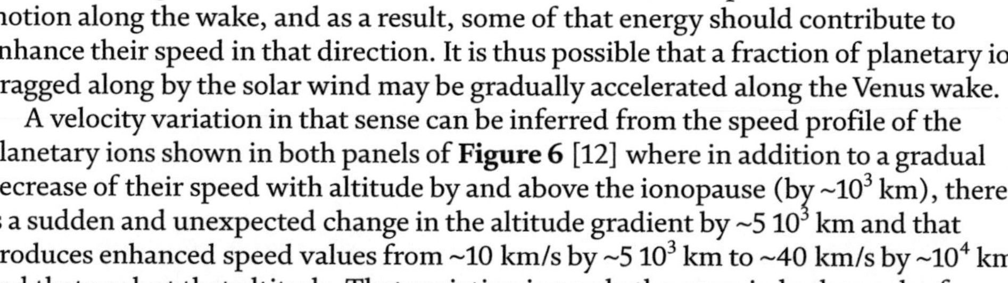

Figure 6.
Measured flow velocities versus VEX altitude for solar wind H+ ions, and ionospheric H+ and O+ ions. The curve marked v_{esc} illustrates escape velocity versus altitude above Venus. The data points represent average values in 50 km altitude intervals sampled within Y = +0.5 of the dawn-dusk Meridian (left panel) and of the noon-midnight Meridian (right panel). Regions and boundaries are marked on the right-hand side as the I-sphere (the ionopause (IP), and the ionosheath (IMB). From Lundin et al. [12].

maintained, and thus a fraction of the energy released will be provided to the particle population that preserves that motion. Following the shape of a vortex structure downstream from Venus that is similar to that indicated in the equivalent form past an object in **Figure 5**, we can suggest a shape that by a distance of two planetary radii along the Venus wake, the thickness of the vortex has decreased significantly. The outcome of that geometry is that when the thickness of the corkscrew flow has decreased, the speed of the particles has to be larger so that their kinetic energy density integrated over the area of the cross section is maintained. Such variation is comparable to that expected from the shape of the corkscrew flow in **Figure 5** where larger speed values are expected where the cross section of the vortex structure is smaller. Thus, the data points in **Figure 6** at low altitudes refer to the gradual motion of the spacecraft through a wide vortex in a region close to Venus and that at higher altitudes apply to the region where the cross section of the vortex is smaller and hence the flow speeds are larger. An implication also consistent with the shape of a corkscrew flow shape in **Figure 5** is the abrupt ending of the speed profile by ~1.5 10^4 km altitude in **Figure 6** and that is not related to any drastic change in the density profile since comparable density values were measured above and below that altitude. Instead, the abrupt ending of the speed profile at that altitude may imply the exit of the spacecraft from the corkscrew flow region along its trajectory. In summary, vortices along the Venus wake seem to be confined within a region whose cross section decreases with distance downstream from Venus and with a population of accelerated planetary ions.

4. Velocity vectors unrelated to the J x B force

Unrelated to the acceleration of planetary O+ ions through the decreasing cross section of vortices along the Venus wake, there are other mechanisms that should also contribute to produce that effect. Most notable is the participation of magnetic fields through the Lorentz **J x B** forces. In order to explore that contribution, a review of the

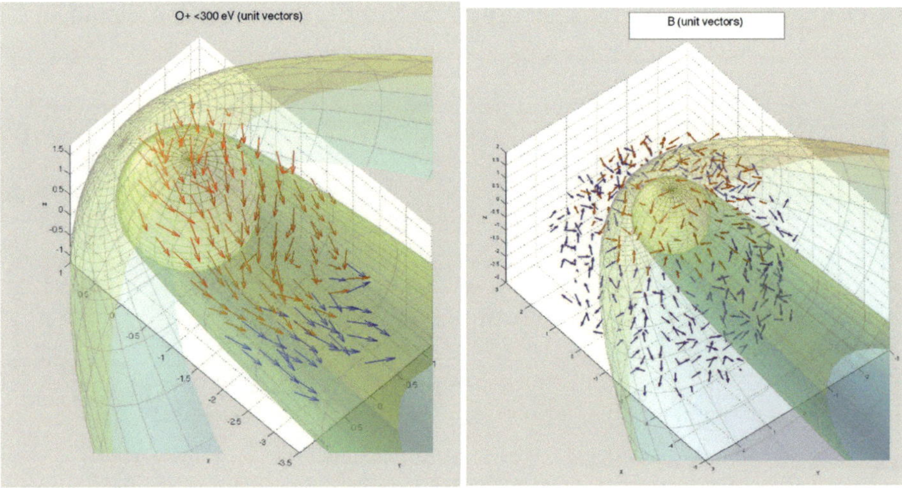

Figure 7.
Average values of the velocity vectors of the planetary O+ ions (left) and the magnetic field vectors (right) measured with the VEX spacecraft along the Venus wake. The direction of the velocity vectors exhibits an orderly configuration while the magnetic field vectors become erratic. Red and blue arrows correspond to Z > 0 and Z < 0 locations.

available VEX data was conducted to examine the manner in which that force may influence particle motion in the wake and that led to the results presented in **Figure 7**. The left panel describes the average values that the velocity vectors of the planetary O+ ions maintain across the wake. Their orderly configuration exhibits a well-organized speed distribution along the wake with some of them displaying a common deflection toward the Y axis. That pattern is entirely different from the erratic distribution of the magnetic field vector orientation measured over many orbits and that is presented in the right panel of that figure. The varying direction of the magnetic field vectors differs from the well-organized orientation of the velocity vectors of the O+ ions and thus provides information that their acceleration is not solely dependent on the **J x B** forces. Other forces mostly derived from wave-particle interactions are necessary to justify the organized distribution of the velocity vectors of the O+ ions, which follow that of the solar wind ions. The data presented here thus provide an important source of information regarding the orderly direction of the velocity vectors of the planetary O+ ion population along the tail and that is unrelated to the **J x B** forces.

5. Conclusions

From the data analysis made in a representative number of VEX orbits, it has been possible to identify the relative position of vortex structures detected before and in the vicinity of a minimum solar cycle with an indication that they occur closer to Venus along its wake under such conditions. This variation was inferred by considering either a small number of VEX orbits or a more extended sample including those obtained across most of the operation time of that spacecraft. A suitable shape of the vortex structures along the Venus wake was also inferred from the observed decrease of its thickness with the downstream distance and that is similar to that of corkscrew flows in fluid dynamic problems. An implication of their expected smaller cross section along the Venus wake is that energy is provided to particles that stream in that direction, thus leading to their gradual acceleration.

The distribution of planetary ions through the Venus wake is arranged by the combined participation of wave-particle interactions on their large-scale gyrotropic trajectories [13]. Their effects should be substantial [14–16] since they will allow the magnetic field forces to be influenced by oscillations and fluctuations of the magnetic field that has draped around the Venus ionosphere and at the same time to the convected electric field of the solar wind. Measurements conducted with the Mariner 5, the PVO, and the VEX spacecraft have contributed with observations of notable magnetic turbulence that is suitable to support that view [17, 18] and that are required to account for the transport of momentum between the solar wind and the ionospheric plasma. Similar conditions should also be applicable to the solar wind/Mars ionosphere interface where there are observations of plasma heating across the flanks of a velocity shear [19], and also that magnetic field wave oscillations have been reported from the MAVEN measurements [20].

Acknowledgements

We wish to thank Gilberto A. Casillas for technical work provided. Financial support was available from the UNAM-IN108814-3 Project.

Author details

Hector Pérez-de-Tejada[1*] and R. Lundin[2]

1 Institute of Geophysics, UNAM, Mexico

2 Swedish Space Research Institute, Kiruna, Sweden

*Address all correspondence to: hectorperezdetejada@gmail.com

References

[1] Pérez-de-Tejada H et al. Plasma vortex in the Venus wake. Eos. 1982;**63**(18):368

[2] Pérez-de-Tejada H, Lundin R, Intriligator D. Plasma vortices in planetary wakes. In: Olmo G, editor. Chapter 13 in: Open Questions in Cosmology. London, UK: Intech; 2012

[3] Lundin R et al. A large scale vortex in the Venus plasma tail and its fluid dynamic interpretation. Geophysical Research Letters. 2013;**40**(7):273

[4] Miller K, Whitten R. Ion dynamics in the Venus ionosphere. Space Science Reviews. 1991;**55**:165

[5] Phillips J, McComas D. The magnetosheath and magnetotail of Venus. Space Sciences Reviews. 1991;**55**:1. (Fig. 35)

[6] Brace L, et al. Holes in the nightside ionosphere of Venus. Journal of Geophysical Research. 1982;**87**:199

[7] Pérez-de-Tejada H et al. Measurement of plasma channels in the Venus wake. Icarus. 2019;**321**:1026-1037

[8] Delva M et al. Proton cyclotron waves in the solar wind at Venus. Journal of Geophysical Research. 2008;**113**:E00B06

[9] Pérez-de-Tejada H, Lundin R. Solar cycle variations in the position of vortex structures in the Venus wake. In: Bouvaquia, editor. Chapter 3 in: Solar Planets and Exoplanets. London, UK: INTECH-OPEN; 2021. DOI: 10.5772/96710

[10] Pérez-de-Tejada H, Lundin R. Vortex dynamics in the wake of planetary ionospheres. In: Vortex Dynamics – From Physical to Mathematical Aspects. London, UK: IntechOpen; 2022. DOI: 10.5772/IntechOpen.101352

[11] Romanov SA et al. Interaction of the solar wind with Venus. Cosmic Research. 1979;**16**:603. (Fig. 5)

[12] Lundin R et al. Ion flow and momentum transfer in the Venus plasma environment. Icarus. 2011;**215**:7

[13] Luhmann J et al. A comparison of induced magnetotails of planetary bodies; Venus, Mars, Titan. Journal of Geophysical Research. 1991;**96**:1199

[14] Shapiro V, Shevchenko V. Astrophysical plasma turbulences. Astrophys and Space Science Reviews. 1968;**6**:427

[15] Szego K et al. Physical processes in the plasma mantle of Venus. Geophysical Research Letters. 1991;**18**(12):2305

[16] Dobe Z et al. Interaction of the solar wind with unmagnetized planets. Physical Review Letters. 1999;**83**(2):260-263

[17] Bridge A et al. Plasma and magnetic fields observed near Venus. Science. 1967;**158**:1669-1673. (see Fig. 2)

[18] Vörös Z et al. Intermittent turbulence, noisy fluctuations and wavy structures in the Venusian magnetosheath and wake. Journal of Geophysical Research. 2008;**113**:ED0B21, 200. DOI: 1029/2008JE003159

[19] Dubinin E et al. Plasma characteristics of the boundary layer in the Martian magnetosphere. Journal of Geophysical Research. 1996;**101**:27061

[20] Ruhunusiri S et al. MAVEN observations of partially developed Kelvin-Helmholtz vortices at Mars. Geophysical Research Letters. 2016. DOI: 10.1002/2016GL068926

Section 4

Plasma Microwave Interactions

Chapter 6

Design and Analysis of High Power RF Window

Mukesh Kumar Alaria

Abstract

Gyrotron has proven to be an efficient source for RF generation at high power and high frequency level. Gyrotron has very significant and specific role to play, particularly, in plasma, material, spectroscopy and energy research. The chapter covers the selection of dielectric materials and power handling capability of RF windows. The design analysis of RF window with window disk thickness and diameter are optimized considering the minimum return loss and the minimum insertion loss by using CST microwave studio. The advantage of the double disc window cooling is most effective. In the 42 GHz window design, double disc winnow are using and the temperature on the disc of RF window does not exceed 100 degree C and found in safe limit. The design of RF window depends on the dielectric characteristics like dielectric constant, loss tangent, permittivity, etc., of the window materials. The dielectric permittivity and loss tangent of window material affect the absorption and transmission of RF power. The thermal analyses are performed by using finite element analysis code ANSYS code on the basis of electrical design of RF window.

Keywords: Gyrotron, loss tangent, reflection, transmission, Gaussian beam, RF losses

1. Introduction

At present the Gyrotron is used almost in every plasma fusion machines as a high power millimeter wave source for electron cyclotron resonance heating (ECRH). RF window is a very critical component of a high power Gyrotron and used on the output section of the device for the extraction of RF power from vacuum to external pressurized atmospheric environment [1–5].

In high power microwave devices, RF window is also a critical component and used for extraction of RF power from the Gyrotron. RF window is used on the output section of the device for the propagation of RF power from the ultrahigh vacuum environment to the external normal pressure environment for the effective extraction of RF power. Thus the careful design of this component is essential for the successful operation of the device. The window disc dimensions are optimized in the design of a RF window taking into account the minimum power reflection and the maximum power handling capability.

RF loss in the RF window directly depends on the dielectric properties of the materials. Obviously, the dielectric properties should be stable in a wide range of temperatures so that the window performance remains stable during the heating. RF window must be fabricated from a low loss material which should be also suitable for the ultra-high vacuum application. Thus, the loss tangent of the RF window material should be minimum for the least absorption of RF power.

The material selection is very important for a Gyrotron window operating at high power and high frequencies. For a high power Gyrotron the advanced materials such as sapphire, chemical vapor deposited (CVD) diamond, silicon nitride composite, BeO, Au-doped silicon etc., have to be used [6–9]. CVD diamond is the first choice for the MW power Gyrotron due to its excellent dielectric and mechanical properties.

The electrical and thermal designs of the RF windows for 42 GHz Gyrotrons are presented in this Chapter. It is of interest to mention that gyrotrons operating at these two frequencies are considered for study as both of these gyrotrons have been developments at CEERI, Pilani under various national projects. The various types of RF windows used in a Gyrotron are discussed in Section 2. The electrical and thermal designs of RF window for 42 GHz Gyrotron is described in Section 3.

2. Gyrotron RF window

Gyrotron window is required for transmitting significant amount of microwave energy without exhibiting thermal runway. The design requirement of RF windows is minimum return loss, minimum insertion loss, power handling capacity and broad bandwidth. Different dielectric material such as, boron nitride, silicon nitride, sapphire and PACVD diamond are used in a design of RF window. Commonly, boron nitride, silicon nitride and sapphire are used in low power and short pulse Gyrotron window. The combination of low millimeter wave losses and excellent thermal conductivity with good mechanical properties makes artificial plasma assisted chemical vapor deposition (PACVD) diamond a compelling choice for high power, high frequency long pulse Gyrotron window. The diamond window gives a solution for the window problem which has been regarded as the most serious issue on the development of high power long pulse Gyrotron [10–16].

In double disc window, two discs are required per barrier with a spacing of approximately one wave length and some power is absorbed by dielectric cooling liquid, which are disadvantages of this types of window. Due to moderate RF power double discs sapphire window has been selected for use in 42 GHz, 200 kW Gyrotron and presented in Section 3. In single disc window, low loss tangent dielectric materials such as boron nitride, silicon nitride, sapphire and plasma assisted chemical vapor deposition (PACVD) diamond are used. The disc is built into an assembly in which two Inconel tubes are bonded on both sides of the plate to provide vacuum shielding and water cooling to the edge of the disc, leaving an effective window aperture of 85 mm. The PACVD diamond window can be used for 1 MW Gyrotron due to its high thermal conductivity and low dielectric loss. Diamond grown by chemical vapor deposition (CVD) is becoming a viable window material [17]. Window must be able to withstand the thermally induced stresses caused by the temperature rise of the window as well as the mechanical stress caused by atmospheric and coolant pressure [18]. In the Gyrotron, different types of windows according to the various cooling schemes are used for the extraction of RF power. Generally, windows with (a) multiple dielectrics with distributed cooling (b) double discs with surface cooling by liquid (c) single disc with surface cooling by gas and (d) single disc with edge cooling are used in Gyrotrons. Each of these windows possesses its own advantages and disadvantages. Basic aspects of gyrotron window are discussed in sub-Section 2.1. Further, the selection of dielectric materials for Gyrotron window is presented in sub-Section 2.2 and the power handling capability of RF window in sub-section Section 2.3, respectively.

2.1 Basic aspect

Ultimately, the high power limits for all RF window is a result of heating caused by absorbed microwave energy in the window. Even though the fraction of energy absorbed is quite small, of the order of 0.1% for typical materials of interest, the large amounts of transmitted power means that even a small fraction can result in significant heating leading to thermal stress and thermal runaway. In the heated RF window structure thermal energy generation is initiated by the absorbed microwave energy [19]. This energy is removed in steady state operation by thermal conduction and force convection. The conduction temperature profile is important because it ultimately leads to failure of the window when the power is high. The temperature gradients required to drive thermal energy out of the window cause different areas of the window. As window heat up, these stresses would reach the ultimate strength for the given material and surface of geometry and the window would fail. Face cooled window is not as susceptible to thermal runaway effects as edge cooled window.

The thickness (d) of window disc is chosen so that the reflection is minimized and it may be initially estimated through the expression given as [20]

$$d = \frac{N\lambda_0}{2\sqrt{\varepsilon_r'}} \tag{1}$$

where N is integer and λ_0 is the free-space wavelength, the maximum transmission occurs at the series of frequencies f_i for which

$$N_i\lambda_i = 2d\sqrt{\varepsilon_r'} \tag{2}$$

where suffix i signifies the various integers. A possible solution for broadband window is Brewster window. Brewster window is used in multi frequency Gyrotron. Window reflections are the main problem in the design of Brewster window. The transmission band of Brewster window is large and the disk in the window is tilted with a specific angle. The angle between the normal to the window disk and the RF propagation axis is chosen according to Eq. (3).

$$\theta_{Brewster} = arc\ tan\sqrt{\varepsilon_r'} \tag{3}$$

The value of disc thickness (d') in case of TE_{mn} mode for low power window can be obtained as:

$$d' = \frac{N\pi}{\sqrt{\left(2\pi f\sqrt{\frac{\varepsilon_r'}{c}}\right)^2 - \left(\frac{\chi_{mn}'}{R_{win}}\right)^2}} \tag{4}$$

where f is frequency, R_{win} is window radius, c is velocity of light and χ_{mn}' is the nth root of the derivative of mth order first kind Bessel function. Based on this formulation

Materials	At 120 GHz $d = \frac{4\lambda_0}{2\sqrt{\varepsilon_r'}}$	At 170 GHz $d = \frac{5\lambda_0}{2\sqrt{\varepsilon_r'}}$	Permittivity (ε_r')
Boron Nitrate (BN)	2.3 mm	2.03 mm	4.7
CVD	2.1 mm	1.85 mm	5.67
Sapphire	1.63 mm	1.44 mm	9.4

Table 1.
Disc thickness for different frequency Gyrotron window.

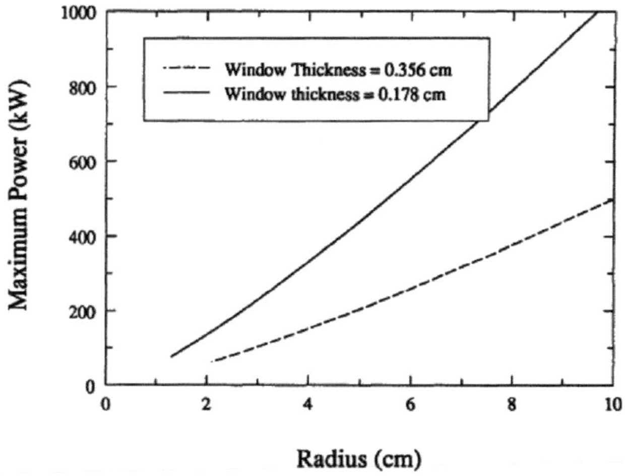

Figure 1.
Effect of window radius on maximum power level of window.

the range of disc and other design parameters considered. It is interest to mention that Eq. (4) can be easily transformed as Eq. (1) for gyrotron window due to insignificant value of second term of the denominator of Eq. (4). The initial disc thicknesses (d) of high frequency window, obtained from eq. (1), for various high power Gyrotrons are given in the **Table 1**.

The effect of varying the window radius is illustrated in **Figure 1**. With the larger window the power handling capacity increases. However, as increase in the window thickness raises the total heat generated in the window and thereby is detrimental to the window thermal performance. It shows that the maximum acceptable power level decreases when the window thickness is increased.

2.2 Selection of window materials

The important aspect of high power Gyrotron window design is the dielectric properties of the window materials. Since the ceramic comes in the path of propagating microwave, the dielectric properties of the ceramic material have strong influence on the electrical design of window. Desired dielectric properties of a window material are low relative permittivity (ε_r') and low loss tangent (tan δ). If all else have been optimized, a lower relative permittivity (ε_r') gives a better bandwidth response [21].

Alumina is the predominant material for high power window applications. Beryllia, though better in electrical and thermal properties than alumina, is seldom used. Beryllia is not mechanically strong, and the potential danger in handling such a toxic material is also a definite drawback. Sapphire has been the best material for a microwave window because of its high strength, low absorbed power and good tolerance of radiation damage. Sapphire is a single crystal material because it is hardest material and has highest melting point of any materials that is commonly available in the industry. Sapphire materials are widely used because its transparency, superior mechanical strength, chemical and scratch resistance and the fact that it can be relatively easily developed as a grown crystal. Sapphire also has the potential for much lower microwave losses as a result of its large reduction in loss tangent at cryogenic temperatures, where it also becomes somewhat stronger.

The effect of poor thermal conductivity was identifiable from the extremely high temperature gradients (up to 600°C) attained immediately prior to failure, but they can handle up to 100 kW power. The dielectric properties should be stable in wide range of temperature so that the window performance being stable during the heating. Due to the very good thermal conductivity and very weak dependency of the dielectric parameters on temperature, diamond is selected for window design for high power and high frequency Gyrotron which is capable for long pulse operation. The diamond window gives a solution for the window problem which has been regarded as the most serious issue on the development of high power long pulse Gyrotron. For the low power Gyrotron the best materials like sapphire and boron nitride are used due to the low cost and easy availability.

The strength and thermal conductivity of PACVD has made it the leading candidate to replace sapphire in RF window applications. The edge cooled window is a major advantage of avoiding complex flow passages in the microwave channel. High power limit for all vacuum windows is a result of heating caused by absorbed microwave energy in the window elements themselves. Even though the fraction of energy absorbed is quite small, on the order of 0.1% for typical materials of interest, the large amounts of transmitted power means that even a small fraction can result in significant heating leading to coolant boiling, thermal stress failure, or thermal runaway. The small but significant absorption is due to a complex component of the dielectric constant, indicating that some polarized or charged constituents of the material move in phase with the applied electric field at the frequency of the microwave beam.

Dielectric materials such as sapphire and boron nitride (BN) are used in RF window for low power gyrotron due to its low failure resistance and low RF power capacity of material, while PACVD diamond is used in high power and long pulse gyrotron due to its high failure resistance and high power capacity of material. The expressions of these two empirical parameters, namely failure resistance *(R')* and power transmission capacity *(P_T)* for edge cooled windows [20]. These parameters are defined as follows:

$$R' = \frac{k\sigma_B(1-\upsilon)}{E\alpha} \tag{5}$$

where k is thermal conductivity, σ_B is bending strength, ν is Poissons number, E is Young's modulus and α is thermal expansion coefficient. Further

$$P_T = \frac{R'\rho c_p}{(1+\varepsilon_r')tan\delta} \tag{6}$$

where ρ is density, c_p is specific heat.

Using the material parameters and beam profiles, irradiation tests with a potential new material for simple water-edge-cooled single disc window, diamond is attractive due to its good mechanical properties, modest dielectric constant, relatively low loss and excellent thermal conductivity. In the temperature range of 200–370 K the loss tangent and the permittivity of diamond are practically constant. The current CVD capabilities have allowed for tests with diamond discs of up to 100 mm diameter and 2.5 mm thickness for megawatt Gyrotrons.

2.3 Power handling capability

RF windows are generally preferred and used for high power microwave tubes due to their higher handling RF power capacity. Performances of microwave tubes are limited by a number of factors; these are heat dissipation, voltage breakdown, and window failure and multipactor phenomena. The maximum power obtainable from a microwave tube is often determined by the power handling capability of the output window. The output RF power is propagated out through waveguide or coaxial line windows. The power handling criteria is rather a check for the diameter and indirectly sets a limit to the useful diameter.

The power profile in the output window can take a variety of shapes. The most desirable profile is Gaussian distribution which simplifies coupling of the RF into a waveguide for transport to the plasma. The failure of RF windows may happen in high power microwave tubes due to the following two reasons. (i) The ceramic disc in window is bombarded by high energy electrons or ions and gives rise to multipactor phenomena that causes discharge on the surface of the ceramic disc. As a result, the ceramic disc is perforated or cracked. And, (ii) The heat losses generates excessive heat in the ceramic disc and the temperature profile of the ceramic disc rises up too high that it cracks developed because the tension caused by excessive heat expansion is too strong to be endured by the material. The most intractable problem for a window used to transmit high power continues wave (CW) results from the excessive heat generated by RF losses in the dielectric disc. Although there are several causes that may generate heat in dielectric disc, the above-mentioned two kinds, that is, electron bombardment and RF losses are the most dominating heat sources in the dielectric disc in a window used to transmit high power CW, and they must be taken into account seriously. Cylindrical TE mode window is comprehensively used in high power microwave tubes. It consists of a thin dielectric disc. When microwave passes through the dielectric disc, charge polarization would be induced and thus yielding RF loss causes the thermal heat in the window material. Because the heat dispersed from the interface of the disc with vacuum or air is too little to be considered, it is transmitted only by conduction to the edge and dispersed through metal window frame and cooling fluid, thus a thermal equilibrium is reached. The edge of the dielectric disc keeps on constant temperature under well cooling condition. The temperature in the centre of dielectric disc is the highest when the thermal equilibrium is reached. If the temperature difference between the center and the edge surpasses a certain limit, the dielectric disc may crack. Assuming the temperature difference between the centre and the edge of the dielectric disc to be T_m, then one can write:

$$T_m = 0.725\frac{f\varepsilon_0\varepsilon_r''P_0Z_d}{h} \tag{7}$$

where h is the heat conductivity coefficient, f is frequency, P_0 is passing power, ε_0 is permittivity in the free space medium, ε_r'' is loss factor in the dielectric medium and Z_d is the characteristic impedance of $TE_{0,3}$ mode in dielectric disc and expressed as follows:

$$Z_d = \frac{\sqrt{\mu_0/(\varepsilon_0\varepsilon_r')}}{\sqrt{1-\left(\frac{\lambda_0}{\lambda_{c\ 03}}\right)^2}} \tag{8}$$

where μ_0 is permeability in free space medium, λ_0 is free space wavelength and λ_c is cutoff wavelength. Therefore, a formula of average power capacity (P_{Total}) of cylindrical TE_{mn} mode window can be acquired as follows:

$$P_{Total} = 8.67\ \frac{h\Delta T_m}{\omega\varepsilon_0\varepsilon_r''Z_d} \tag{9}$$

where ΔT_m is the endurable maximal temperature difference between the centre and the edge of a dielectric disc. The temperature increment should be as large as possible to be endured by ceramic disc. Moreover, it has been concluded in that the maximal endurable temperature difference between the centre and the edge of dielectric disc made of ceramic is about 100 K.

i. **RF Power loss in waveguide:** For TE cylindrical waveguide mode profile function $f(z) = e^{-ik_z z}$ of a forward propagating wave, the power flow in the waveguide is given by:

$$\mathrm{P_{rf}} = \frac{1}{2\mu_0}\int |\mathrm{E}\times\mathrm{B}^*|\mathrm{da} = \frac{\pi k_z}{4\mu_0\omega k_\perp^2}\left(\chi_{mn}'^2 - P_0^2\right) J_m^2(\chi_{mn}') E_0^2 \tag{10}$$

where E and B^* are the electric and the complex conjugate of magnetic field respectively, E_0 is amplitude of electric field, P_0 is passing power and k_z, $k_\perp$ are the axial wave number and the transverse wave numbers, respectively. Further, the ratio of ohmic loss to RF loss in the window is given by

$$\frac{\mathrm{P_{ohm}}}{\mathrm{P_{rf}}} = \sqrt{\frac{2}{\mu_0\omega\sigma}\frac{\mathrm{k}_\perp^4}{\pi \mathrm{k_z}}}\ \frac{1}{\left(\chi_{\mathrm{mn}}'^2 - \mathrm{P_0}^2\right)}\left(1 + \frac{\mathrm{m^2k_z^2}}{\chi_{\mathrm{mn}}'^2\mathrm{k}_\perp^2}\right) \tag{11}$$

TE_{0n} mode: The RF power generated in a TE_{mn} circular waveguide mode and incident on a window disc of radius (R) can be found by using Poynting theorem. For azimuthally symmetric modes with $m = 0$, we found power density distribution, S_z (r) as

$$S_z(r) = \frac{P_{Total}\ \chi_{mn}'^2}{\pi R^2}\left[\frac{J_1^2\left(\frac{\chi_{0\ nr}}{R}\right)}{J_0^2(\chi_{0n})}\right] \tag{12}$$

where P_{Total} is the total power.

TEM_{00} Gaussian beam mode: For a fundamental symmetric Gaussian beam, the power density distribution is given by

$$S_z(r) = P_{Total}\ \frac{2}{\pi w^2} e^{-2\left(\frac{r}{w}\right)^2} \tag{13}$$

where w is the Gaussian beam radius, defined as the value of r at which the field amplitude is 1/e times its on axis value. Thus the radial power absorption in the disc can be modeled by:

$$P(r) = A\ S_z(r) \tag{14}$$

where A is the power absorption coefficient and is given as:

$$\mathrm{A} = 2\,\alpha\,\mathrm{l} \tag{15}$$

where α is the attenuation due to a dielectric region of length l in a waveguide.

ii. **RF loss in ceramic window:** The RF loss in the ceramic window may be written as

$$\alpha_0 = \frac{2\pi f}{c} \sqrt{\varepsilon_r'}\, tan\delta\, S \tag{16}$$

where α_0 is the absorption loss. Further, S is the enhancement factor in the absorption due to the fact that the windows are acting as Fabry-Perot. S is close to unity when the window is made anti-resonant (which requires a tolerance on the window thickness). For the case of resonant power transmission, which is considered here, S is given as:

$$S = \frac{(1 + \varepsilon_r')}{2\sqrt{\varepsilon_r'}} \tag{17}$$

On the basis of eq. (10)–(17), an algorithm has been found out and computer program in MATLAB has been developed. **Figure 2** shows the power distribution $S_z(r)$ and loss on the disc for TE mode.

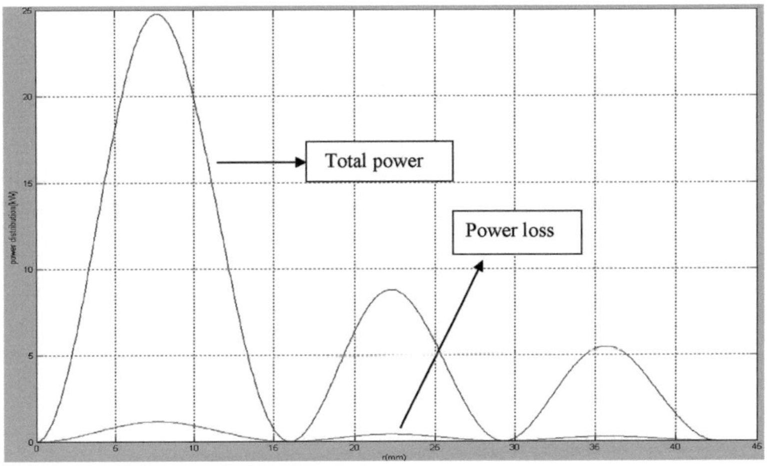

Figure 2.
Power distribution $S_z(r)$ vs. disc of radius for TE mode.

Materials	Enhancement factor (S)	RF losses in ceramic (α_O) At 42 GHz	RF losses in ceramic (α_O) At 120 GHz
Sapphire	1.7	0.90	2.61
CVD	1.4	0.05	0.16
BN	1.3	2.83	8.10
Alumina	1.7	0.46	1.31

Table 2.
RF losses in the dielectric for different materials.

S.No.	Input parameters	Value	Power distribution in edges of disc $S_z(r)$
1.	Output power (P_O)	200 kW	78 kW
2.	Mode (TE $_{mn}$)	$TE_{0,3}$	
3.	Waveguide Radius (a)	42.5 mm	
4.	Frequency (f)	42 GHz	

Table 3.
RF power distribution in the sapphire disc window.

Input parameters	Value	Power handling capacity (P_T)	Remarks
Radius (a)	42.5 mm	194 KW	(ΔT_m) Temperature difference is high power capacity (P_T) will be increased.
Passing Power (P_O)	200 kW		
Frequency (f)	42 GHz		
Temperature difference (ΔT_m)	107 K		
Characteristic impedance of $TE_{0,3}$ mode (Z_d)	135 ohm		

Table 4.
Power handling capacity (P_T) of sapphire window.

Table 2 gives the enhancement factors and RF losses in the ceramics for different dielectric materials. **Table 3** gives the power density distribution in the edges of the sapphire disc, calculated. In the same way, **Table 4** gives the power handling capacity of sapphire window for 42 GHz Gyrotron at high temperature estimated.

3. Design of RF window for 42 GHz, 200 kW Gyrotron

The design of face cooled double discs window for 42 GHz, 200 kW Gyrotron has been carried out using the CST microwave studio (CST-MS). The window disc thickness and diameter are optimized considering the minimum return loss and the minimum insertion loss. The dielectric properties are used for the S-parameter estimation in CST simulation and the mechanical and thermal properties are used for the thermal and structural analysis of RF window in ANSYS simulation. In this section, the electrical design of RF window for 42 GHz, 200 kW.

3.1 Electrical design of RF window for 42 GHz Gyrotron

In 42 GHz, 200 kW Gyrotron double discs, each of diameter 85 mm and thickness 3.2 mm sapphire window and spacing of discs 2.7 mm have been optimized in the simulation where sapphire discs are face cooled by Coolant FC-75. These values of window disc diameter, thickness and spacing are optimized considering the minimum return loss (≤ -20 dB) and the minimum insertion loss (≤ 0.1 dB) by using CST microwave studio [22]. **Figure 3** shows the schematic diagram of double disc faced cooled sapphire window for 42 GHz Gyrotron. The material of disc is sapphire and dielectric constant (ε_r') of the sapphire is 9.41, loss tangent is 5.4×10^{-5} while dielectric constant of FC-75 is 1.8 and loss tangent ($tan\ \delta$) is 26×10^{-4}.

The design flow chart of high frequency and high power Gyrotron window is shown in **Figure 4**. The performance of the S-parameters for 42 GHz double discs faced cooled window with respect to frequency is shown in **Figure 5**. The S parameter simulations for the optimized RF window show very small reflection and absorption in the RF power. The disc thickness of window is small shows some advantages such as proper metallization for better mechanical strength and brazing. The reflection and transmission of the window are independent from the diameter as the wavelength is quite small as compared to the ceramic disc outer diameter. The return loss performance with respect to frequency for various RF window disc thicknesses of sapphire material is shown in **Figure 6**.

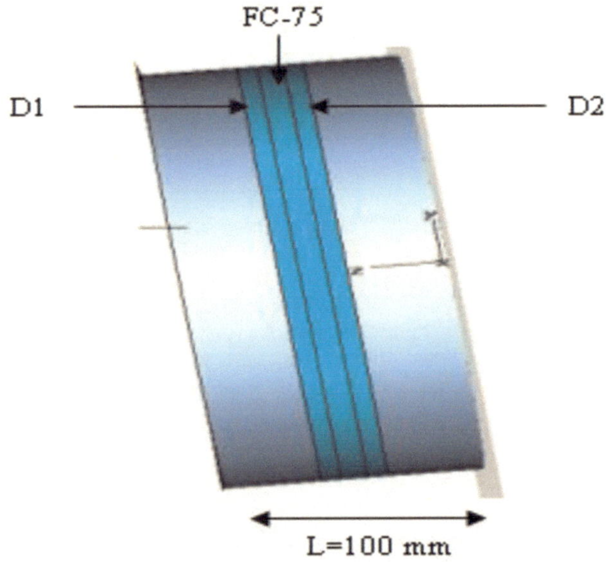

Figure 3.
Faced cooled double disc window for 42 GHz Gyrotron.

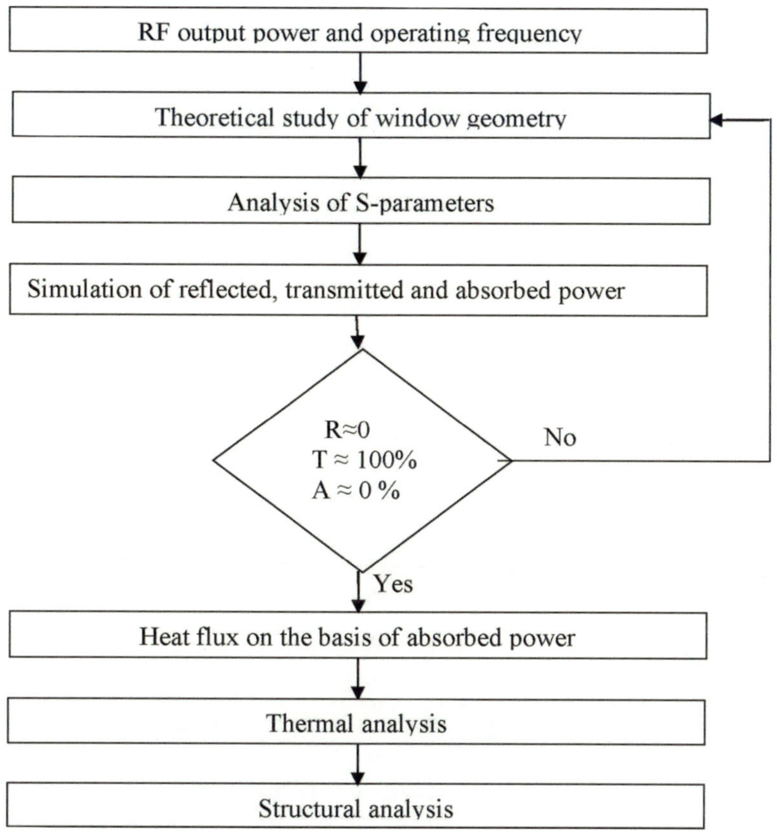

Figure 4.
Design flow chart for any Gyrotron window.

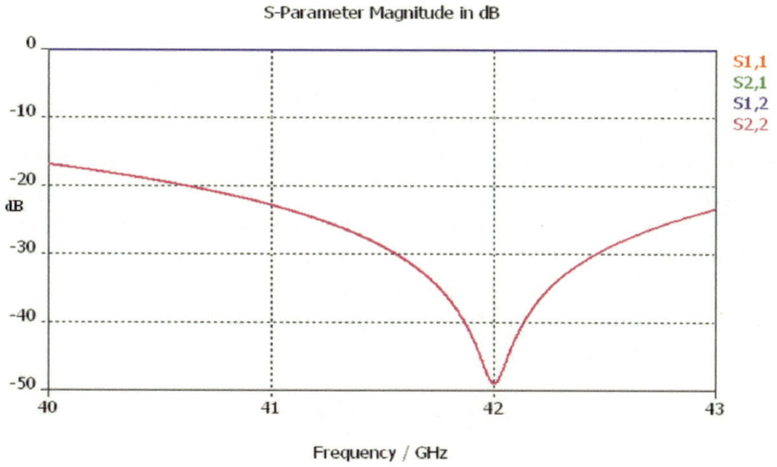

Figure 5.
Return loss performance of double disc sapphire window for 42 GHz Gyrotron.

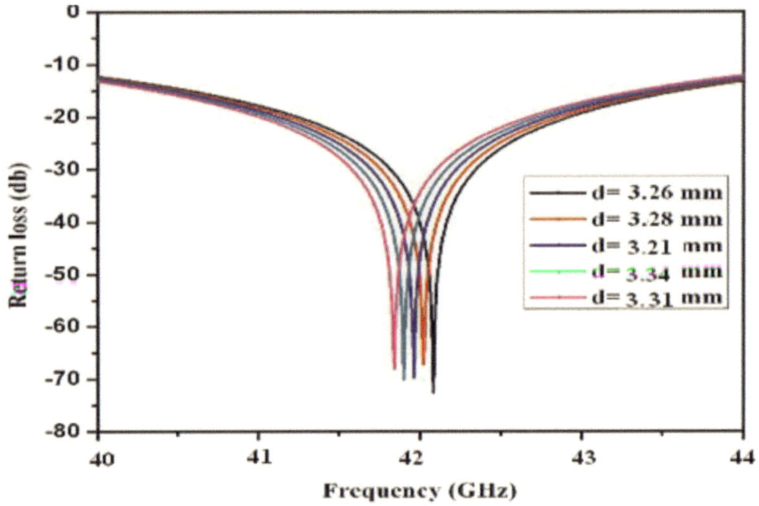

Figure 6.
Return loss performance of different thickness for sapphire disc window.

3.2 Thermal design

On the basis of electrical design of RF window, the thermal and structural analysis are performed by using finite element analysis code ANSYS [23]. The objective of thermal analysis is to assess the temperature distribution, axial stress, radial stress and thermal expansions in the RF window for the 42 GHz Gyrotron during extreme case of operation, that is, at saturation. The return loss and insertion loss of the RF window have been obtained as −49.5 dB and − 0.02 dB, respectively. Using these values of return loss and insertion loss, the reflected, transmitted and the absorbed powers by double sapphire discs can be easily estimated as 0.001%, 99.54% and 0.42%, respectively. The beam-wave interaction simulation results carried out in Gyrotron Lab, CEERI, shows 280 kW RF power generation in the interaction cavity and thus 1176 W RF power is absorbed by the window disc in double discs sapphire window. This absorbed power is distributed all over the disc surface as TE_{03} mode.

A transient thermal analysis is used to determine the temperature distribution as a function of time. The input parameters are heat flux, film coefficient and bulk temperature are using in the analysis. The output parameter is the temperature distributions on RF window. The design of cooling system is chosen so that maximum surface of the window is covered and flow rate of coolant is varying and optimized from 5.0 to 10 lit/min for maximum cooling. The windows boundaries consist of two parallel faces with coolant in between and heat flow across the other face is given as:

$$q = h(T_s - T_b) \tag{18}$$

where h is heat film coefficient, T_s the surface temperature of the disc and T_b bulk temperature of the coolant. The edges of the sapphire disc are assumed to be isolated. The heat flux is given as an input parameter in the thermal simulations. First of all, the heat film coefficient is optimized considering the effective cooling of the window disc. The window geometries are optimized for the effective cooling. Normal water (288 K) is used as a coolant in the thermal simulations of the double discs sapphire window. Temperatures on centre of disc and edge of the disc in double discs window geometry are 77.13°C and 35°C, respectively.

4. Conclusion

The design methodology of RF window for high frequency Gyrotron is presented in this Chapter. Both electrical and thermal analysis for microwave window design has been discussed in this Chapter. At first, the various basic aspects, the selection of window materials and the power handling capability of high power window have been discussed. It is found that Sapphire and BN can be used in moderate power Gyrotron but CVD diamond is must for megawatt gyrotron. In 42 GHz, 200 kW Gyrotron double discs, each of diameter 85 mm and thickness 3.2 mm sapphire window and coolant FC-75 has been optimized in the design using CST microwave studio. In the presented thermal design through ANSYS simulation, the temperature on the disc of RF window does not exceed 100°C and thus within the safe limit. The subsequent thermal stresses also do not deteriorate it. Single disc edge cooled window geometry is used for Gyrotron tube which allows the microwave radiation to exit the tube. The thermal analyses are performed by using finite element analysis code ANSYS code on the basis of electrical design of RF window. The temperature on centre of disc and edge of the disc in single disc window geometry are found as 376 K and 310 K, respectively. The water-edge-cooled CVD diamond window provides a practical technical solution for the transmission of continuous millimeter wave transmission in excess of 1.0 MW.

Author details

Mukesh Kumar Alaria
CSIR-CEERI, Pilani, Rajasthan, India

*Address all correspondence to: mka@ceeri.res.in

References

[1] Nusinovich GS. Review of the theory of mode interaction in gyro devices. IEEE Transactions on Plasma Science. 1999;**27**:313-326

[2] Edgcombe CJ, editor. Gyrotron Oscillators: Their Principles and Practice. London: Taylor & Francis; 1993

[3] Jory H. Gyro-device development and applications. International Electron Devices Meeting Technical Digest. 1981; **11**:182-185

[4] Hirshfield JL, Granatstein VL. Electron cyclotron maser — An historical survey. IEEE Transactions on Microwave Theory and Techniques. 1977;**25**:522-527

[5] Gold SH, Nusinovich GS. Review of high-power microwave source research. The Review of Scientific Instruments. 1997;**68**:3945

[6] Makowski M. ECRF systems for ITER. IEEE Transactions on Plasma Science. 1996;**24**:1023-1032

[7] Thumm M. State- of- the- Art of High Power Gyro-Devices and Free Electron Masers Update 2010. Germany: FZK, KIT; 2011

[8] Thumm M. State-of-the-Art of High Power Gyro-Devices and Free-Electron Masers: Update 1955, FZKA Report 5728. Karlsruhe: Institut for Technische Physik; 1996

[9] Kartikeyan MV, Borie E, Thumm MKA. Gyrotrons-High Power Microwave and Millimeter Wave Technology. Karlsruhe, Germany: Springer; 2004

[10] Gilmour AS Jr. Klystron, Traveling Wave Tubes, Magnetrons, Crossed-Field Amplifiers and Gyrotrons. Boston: Artech House; 2011

[11] Thumm M. MW gyrotron development for fusion plasma applications. Plasma Physics and Controlled Fusion. 2003;**45**: A143-A161

[12] Dumbrajs O, Nusinovich GS. Coaxial gyrotrons: Past, present and future. IEEE Transactions on Plasma Science. 2004; **32**:934-946

[13] Petlin MI. One century of cyclotron radiation. IEEE Transactions on Plasma Science. 1999;**27**:294

[14] Scott AW. Understating Microwaves. New York: John Wiley and Sons; 1993

[15] Dumbrajs O, Liu S. Kinetic theory of electron cyclotron resonance masers with asymmetry of the electron beam in a cavity. IEEE Transactions on Plasma Science. 1992;**20**:126

[16] Nusinovich GS. Introduction to the Physics of Gyrotron. Maryland, USA: JHU; 2004

[17] Singh VVP, Arnold A, Borie E, Braz O, Thumm M. Thermal modeling of edge-cooled single disc gyrotron windows using a one-dimensional finite difference computer code. International Journal of Infrared and Millimeter Waves. 1998;**19**:1451

[18] Alaria MK, Mukerjee P, Das S, Sinha AK. Study of cavity and output window for high power Gyrotron. Journal of Fusion Energy. 2011;**30**(1):89-93

[19] Heidinger R, Dammertz G, Meier A, Thumm M. CVD diamond windows

studied with low and high power millimeter waves. IEEE Transactions on Plasma Science. 2002;**30**:800

[20] Haldeman GS. A Novel Window for Megawatt Gyrotrons [Ph.D Thesis]. USA: MIT; 2001

[21] Chatterjee R. Microwave, Millimetre-Wave and Submillimetre-Wave: Vacuum Electron Devices. New Delhi: Affiliated East-West Press; 1999

[22] CST Microwave Studio Version 11.0, Computer Simulation Technology Darmstadt, Germany.

[23] Ansys Inc. ANSYS Vs 11.1: User Manual. USA: Ansys Inc.; 2010

Chapter 7

Reconfigurable Antennas Based on Plasma Reflectors and Cylindrical Slotted Waveguide

Fatemeh Sadeghikia, Ali Karami Horestani and Mohamed Himdi

Abstract

In this chapter, we focus on the application of plasma structures to realize reconfigurable antennas. Several approaches are presented to dynamically control the beamwidth and radiation gain of circularly polarized helical antennas based on plasma reflectors. Ideas and design principles were discussed and confirmed by full-wave simulations and measurements of realized prototypes. It is shown that plasma reflectors can be effectively used to design reconfigurable helicone antennas with controllable gain and beamwidth. The chapter also presents a reconfigurable slotted antenna using a plasma tube inside the metallic waveguide. It is shown that the radiation pattern of the antenna can be readily reconfigured by changing the state of the plasma column. In short, it is shown that in contrast to conventional methods based on electronic or mechanical devices, reconfigurable antennas based on plasma media benefit from simple and relatively low-cost structures as well as high performance.

Keywords: plasma technology, reconfigurable antenna, beam steering, beamwidth control, helicone antenna

1. Introduction

During past decades, potential applications of reconfigurable antennas in single-input single-output (SISO) and multiple-input multiple-output MIMO communication systems have been widely investigated [1–32]. In both cases, the principal objective is to reduce the complexity of communication systems by obtaining diverse radiation patterns and operating frequencies. For example, beam-steerable antennas can provide multiple functionalities, thus, mitigating the complexity and overall cost of communication systems.

Regarding applications in reconfigurable antennas, partially ionized plasma elements have the potential to be used either as the main radiator of the antenna as a reflector, a parasitic director element, or even as a dielectric to improve the performance of conventional antennas. Various realizations of antennas with reconfigurable radiation characteristics and/or tunable resonance frequencies based on plasma radiating element (s) have been reported in [1, 10–32]. However, due to the finite conductivity of the plasma elements, these antennas generally suffer from low efficiency [12, 33]. This problem may be worse for small plasma antennas on which there are high current densities. It can be seen, however, that since the current density on the reflecting and parasitic elements of an antenna is much smaller than on the main radiator of the antenna, plasma conductors may be good surrogates for metallic reflector structures. The main advantage of using plasma reflectors rather than metallic ones is that plasma reflectors can be switched on or off to toggle between original and modified radiation patterns. This feature is demanded in many applications. For instance, when radiation patterns with an adjustable beamwidth or main direction are needed. In brief, a combination of plasma elements as reflectors and metallic element (s) as the main radiator results in highly efficient reconfigurable antennas. In view of these advantages, this chapter focuses on applications of plasma media as reflectors (rather than the main radiator) for the realization of reconfigurable antennas.

The chapter is organized as follows. In Section 2, the characteristics of the plasma media and the electromagnetic waves propagating in such media are briefly reviewed. In Section 3, the benefits of the application of the plasma reflectors along with conventional metallic antennas are described and two different prototypes, namely, a plasma-cupped-ground helical antenna and a plasma helicone antenna are introduced. For each prototype, the basic structure and the implementation technique are discussed and their radiation characteristics are presented. The study will be concluded in Section 4.

2. Theory of plasma

In order to investigate the effect of plasma media on the radiation characteristics of antennas, some basic properties of the plasma are first presented in this section.

The complex permittivity up of an isotropic plasma at low pressure can be modeled as follows:

$$\frac{\varepsilon_p}{\varepsilon_0} = \left(1 - \frac{\omega_p^2}{\omega^2 + \upsilon^2}\right) - j\left(\frac{\omega_p^2 \upsilon}{\omega(\omega^2 + \upsilon^2)}\right) \tag{1}$$

Based on this relation, the plasma is a dispersive medium with permittivity εp, which is a function of the operating angular frequency ω (rad/s). The permittivity of the plasma is also a function of the electron-neutral collision frequency υ in Hz, and the plasma angular frequency ω_p (rad/s), which is defined as follows [13, 14]:

$$\omega_p = \left(\frac{ne^2}{m\varepsilon_0}\right)^{\frac{1}{2}} \tag{2}$$

where n is the electron density (m^{-3}), e is the electron charge, and m is the electron mass (kg). The electrical conductivity σ is given by [14]:

$$\sigma = \left(\frac{\varepsilon_0 \upsilon \omega_p^2}{\upsilon^2 + \omega^2}\right) - j\left(\frac{\varepsilon_0 \omega \omega_p^2}{\upsilon^2 + \omega^2}\right) \tag{3}$$

This relation thus indicates that the electrical conductivity of the plasma can be controlled at a given operating frequency by varying either the plasma frequency or

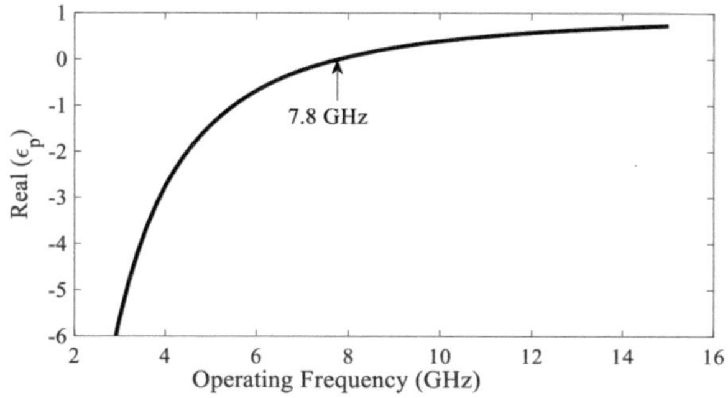

Figure 1.
Variations of the permittivity of the plasma medium versus the frequency for the plasma frequency of 7.8 GHz and the collision frequency of fp/3.

the collision frequency. The loss factor tanδ of the plasma medium can be derived from the ratio of the imaginary part to the real part of the plasma dielectric constant as follows:

$$\tan\delta = \frac{\varepsilon_i}{\varepsilon_r} = \frac{-\left(\frac{\omega_p^2 \nu}{\omega(\omega^2+\nu^2)}\right)}{\left(1 - \frac{\omega_p^2}{\omega^2+\nu^2}\right)} = -\frac{\nu\omega_p^2}{\omega\left(\omega^2 + \nu^2 - \omega_p^2\right)} \tag{4}$$

Based on this relation, plasma loss is a function of plasma parameters and operating frequency.

From Eq. (1), the plasma medium acts as a dielectric with a positive permittivity for incident EM waves with frequencies higher than the plasma frequency, thereby allowing the propagation of the incident wave through the plasma medium. I understand this. In contrast, at frequencies less than the plasma frequency, the plasma shows a negative permittivity, thus prohibiting the propagation of the EM wave. At frequencies much lower than the plasma frequency, the plasma medium can be used as a conductor, but it is not a very good conductor. For instance, in **Figure 1**, variations of the real part of the plasma permittivity versus the operating frequency, for a medium with the plasma frequency f_p = 7.8 GHz and the collision frequency $\upsilon = f_p/3$ are shown. It is observed that at the frequencies below the plasma frequency, the plasma permittivity is negative, thus the plasma reflects the incident wave. However, at frequencies greater than 7.8 GHz, the plasma permittivity varies from 0 to 1, thus the plasma acts as a dielectric. It is also important to note that, since the plasma angular frequency ω_p can be tuned by adjusting some of the characteristics of the plasma, such as its density, the frequency band in which the plasma behaves as a dielectric or a conductor is adjustable.

3. Application of plasma reflectors in the antenna structures

Application of the plasma reflectors along with conventional metallic antennas may provide benefits including steering capability and/or beamwidth control. In general, a variety of methods are available to control the radiation pattern of directional antennas. However, almost all of these methods employ a kind of switching mechanism including electrical, optical, or mechanical switches which require complex and costly feed circuitries [5–7]. Altering the substrate characteristics by using materials such as liquid crystals, ferrites, or plasmas is an almost new alternative method for steering the radiation pattern of the antennas. The unique properties of plasma as the fourth state of the matter have opened a new perspective for the realization of antennas with a reconfigurable radiation pattern or a tunable operating frequency. Therefore, plasma-based reconfigurable/tunable antennas can be considered as cost-efficient and efficient alternatives for currently in-use methods [2, 4, 8, 9]. As shown in the previous section, the plasma medium can be used as a conductor at frequencies much lower than the plasma frequency. An important advantage of plasma-conducting elements compared to metallic ones is that they can be switched on and off and their length can be adjusted electrically. Authors [2, 4, 8, 9, 34, 35] investigated theoretical aspects and different configurations of metal antennas with plasma reflectors for manipulating antenna radiation patterns.

Although various techniques have been introduced to control the beamwidth of antennas, including the use of liquid metal reflectors [36], conformal selective surfaces [37], and varactor switches, antenna beam steering and beamwidth control There has been little research done on the simultaneous execution of Single antenna architecture [38, 39]. However, in these studies, antenna beam steering and beamwidth control increased the cost of hardware and software complexity. It has been shown by authors [40] that a reconfigurable antenna based on plasma reflectors not only may be able to steer the radiation pattern in 3-D space but also can be used for the dynamic control of the antenna beamwidth. Based on these advantages and recent trends in scientific reporting, we believe it is likely that the next decade will see a significant increase in research on plasma-based communication components, including reconfigurable antennas. Therefore, in this section, we present here two prototypes of reconfigurable antennas based on plasma reflectors and analyze their radiation properties. In both cases, a conventional axial-mode helical antenna is used as the primary radiator, and a plasma reflector is used to steer the beam or change the beamwidth of the antenna.

3.1 Plasma-cupped-ground helical antenna

The first part of this section presents the structure of a conventional dish ground axial mode helical antenna and compares its radiation pattern with that of an identical planar ground antenna. In the following, we consider an axial mode helical antenna with a plane ground as the reference antenna. In addition, we study the effect of the height of the dish bottom on the radiation properties of the antenna, including the gain and half-width (HPBW) of the antenna. Part 2 investigates the feasibility of beamwidth control based on plasma elements by replacing the metal dish ground of the helical antenna with a plasma dish ground and numerically comparing the characteristics of the two antennas. Then, the final part of this section describes the realization and experimental verification of the proposed antenna.

3.1.1 Structure and dimensions of dish bottom helical antenna

Various methods have been proposed in the literature to improve the radiation properties of helical antennas. Authors [41–44] considered modifying the ground plane of the antenna to form a cup-shaped ground reflector. Despite its simplicity, this method is very effective in improving the beamwidth and radiation gain of helical antennas. **Figure 2** shows the shape of a conventional dish-bottom helical antenna. The antenna consists of a helical antenna operating in on-axis mode and a dished

DOI: http://dx.doi.org/10.5772/intechopen.108017

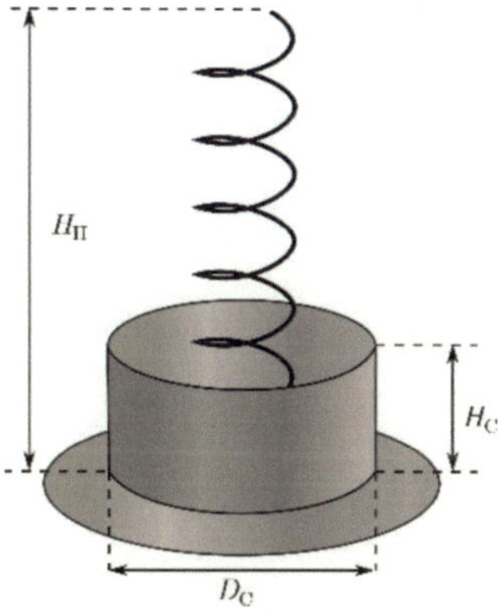

Figure 2.
The structure of a typical cupped ground helical antenna.

metal reflector. It can be shown that the cup-shaped reflector with the height of H_C, and the diameter D_C can improve the radiation gain and the HPBW of the antenna. Note that the achieved gain depends on the dimensions of the cup, including its height and radius. For demonstration, the effect of the height H_C of the cup on the radiation gain and HPBW of the helical antenna is shown in **Figure 3**. The antenna operates at the center frequency of 927 MHz and has an antenna diameter of D_c = 330 mm ($\sim \lambda$). Simulation results are shown for three different numbers with turns N = 6.5, 10, and 13. In this study, the diameter of the helix is D_H = 116 mm, the pitch angle α = 13.6°, and the helix height H_H = 550 mm. The designed helical antenna with a planar ground conductor with a radius D_g = 480 mm demonstrates a simulated radiation gain of 12.7 dBi and the HPBW of 37.5° at the center frequency. The full-wave simulated radiation gain and HPBW of the antennas versus the height of the cupped ground (normalized to the helix height, i.e., H_C/H_H) show that increasing the height H_C of the cupped ground up to around H_C/H_H = 0.07 when N = 6.5 in this study, enhances the gain of the antenna up to 0.5 dB. However, further increasing the H_C, decreases the gain and consequently, increases the HPBW of the antenna. For instance, when H_C/H_H = 0.29 (i.e., H_C = 160 mm for N = 6.5), the radiation gain of the antenna with respect to the reference helical antenna decreases around 0.4 dB, while the HPBW increases around 7.5°. **Figure 4** compares the simulated radiation gain of the reference antenna with that of the cupped ground helical antenna when H_C = 160 mm for N = 6.5.

It can be concluded here that HPBW and the maximum radiation gain of the helical antenna can be adjusted by the height of the cupped ground. The results also show the potential application of the tunable length plasma cup for controlling the radiation

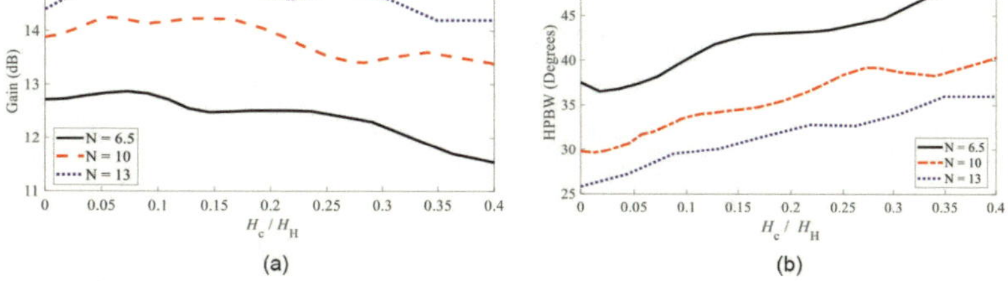

Figure 3.
Simulated Gain and HPBW of the cupped ground helical antenna versus H_C/H_H considering the constant height of the helix (H_H).

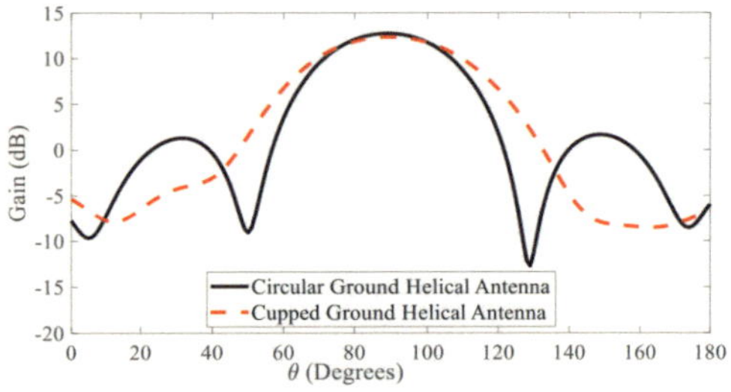

Figure 4.
A comparison between the simulated radiation gain of a conventional helical antenna with a planar ground and the cupped ground helical antenna with H_C = 160 mm.

gain and beamwidth of helical antennas. Thus, in the next section, first, the metallic cup of the helix antenna is replaced with a plasma cup with H_C = 160 mm, and the effects of different configurations of the plasma reflector on the performance of the plasma-based and metallic antenna will be compared.

3.1.2 Plasma reflectors as the cupped ground for a helical antenna

It is worth reminding that if the plasma frequency is sufficiently larger than the operating frequency, the plasma medium demonstrates a relatively high conductivity. Therefore, if the metallic cup in the antenna is replaced with a plasma cup, the efficiency of the antenna with a cup-shaped plasma reflector depends on the plasma frequency. To further investigate this effect, two different plasma frequencies f_{p1} = 3 GHz and f_{p2} = 30 GHz for the plasma-cupped ground in a helical antenna with identical dimensions as the antenna presented in the previous section are studied. In **Figure 5**, the radiation gain of the antenna for the two plasma frequencies f_{p1} = 3 GHz and f_{p2} = 30 GHz is plotted. The results are also compared to the radiation gain of the helical antenna with a metallic planar ground as well as the gain of the antenna with metallic-cupped ground. As shown in this figure, the gain of the first plasma-cupped antenna shows no improvement with respect to the gain of the helical antenna with a planar ground. This is because the conductivity of the plasma cup with the plasma frequency f_{p1} = 3 GHz is not adequate to act as a good reflector. In contrast, for the higher plasma frequency of f_{p2} = 30 GHz, the plasma behaves nearly the same as a metallic cup, leading to improved gain. In short, the simulation results show that the metallic reflector can be replaced with the plasma reflector if the plasma frequency for the plasma reflector is properly chosen.

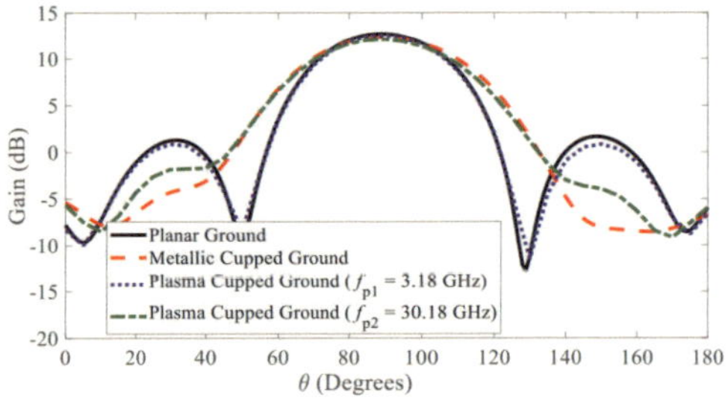

Figure 5.
Simulated radiation gain of a cup-helix antenna with different cup materials compared to a conventional flat-bottom helix antenna.

It is worth noting that achieving a uniform plasma surface is demanding and expensive. However, an array of plasma columns dense enough to form the cup-shaped bottom 10 can be used. For this purpose, taking into account the various types of commercially available plasma tubes. In this study, a compact U-shaped fluorescent lamp (CFL) is used to implement the plasma reflector. An array of 25 CFLs arranged in a circle around the helix is used, as shown in **Figure 6a**. As shown in **Figure 6b**, the actual size of commercial type CFL is considered in this investigation for the geometric scales of the reflectors. In this study, the height of each CFL from the ground plane is 160 mm while its diameter is 14.7 mm leaving a 1.6 mm space gap between the columns of the CFL.

Using this configuration, variations of the number and arrangement of the activated CFL reflectors shape the beam profile of the radiated wave. For the beamwidth adjustment, three configurations are studied in this investigation. The first configuration is called the all-off state, with all CFLs off (), the second configuration with 12 alternate CFLs; on (), and the final configuration with the all-on state. All elements are known to be on ()

For simulation purposes, the plasma frequency and collision frequency of the CFL are assumed to be ω = 14 × 10 rad/s and ν = 400 MHz, respectively. Also, his commercially available CFL tube components are modeled as Pyrex with a dielectric constant of 4.82. and 0.5mm thick. Additionally, inactive CFL tubes are only simulated as empty U-shaped Pyrex tubes. For simulation purposes, a loss factor tanδ = 0.069 at the operating frequency f = 927 MHz is considered. As already mentioned, plasma medium losses are higher than conventional conductors. However, since the plasma element is used as a reflector rather than the antenna's main radiator, the performance of the antenna is not degraded. However, some of the physical parameters of the plasma, such as pressure, can be used to further reduce the effects of plasma leakage [45].

Figure 7 shows the simulated antenna radiation gain for the three configurations. As shown in this figure, a relatively narrow radiated beam, which is around 37°, exactly the same as the radiated beam of the helical antenna with no reflector, belongs to the All-OFF state, while a wider beam and a lower gain are achieved by the 12 alternatively turned-on. Increasing the number of activated CFLs makes the beam

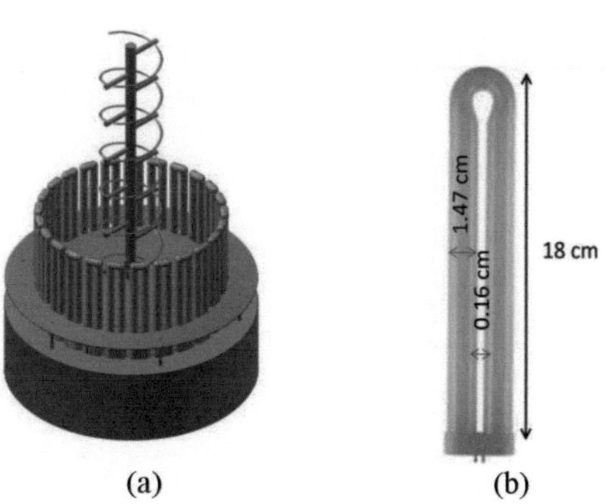

Figure 6.
(a). Geometry configuration of a plasma-cupped-ground helical antenna based on the CFL reflectors. (b). Geometry of the reflective CFL elements.

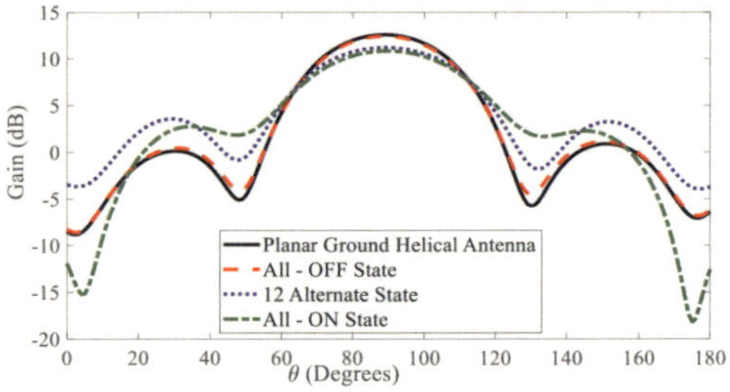

Figure 7.
Simulated radiation gain of the cupped ground helical antenna using plasma reflectors.

even wider and it covers a wider space angle. Therefore, the widest beam, which is around 50.3°, is observed at the direction of the end-fire when all the CFLs are activated and as expected, the gain adversely decreases from 12.5 dBi for the All-OFF state to 9.21 dBi for the All-ON state. In short, as shown in **Figure 8**, due to the diffraction of electromagnetic waves, when a higher number of plasma reflectors is activated, the radiation gain decreases.

Another desirable characteristic of the plasma-cupped-ground helical antenna is that by unsymmetrical activation of a group of neighboring CFLs the main radiation lobe of the antenna can be steered. The radiation patterns of the antenna for two configurations, including the case in which half of the adjacent reflectors (i.e., a half-cylinder) on the right side are turned ON and the CFLs in the left half cylinder are turned OFF, and also the case with its complementary configuration, in which the CFLs in the right half cylinder are turned OFF and those in the left half cylinder are turned ON, are compared and the results are shown in **Figure 9**. The figure clearly shows that the beam of the antenna can be steered by $\pm 6^{\circ}$ electrically by this method. Note that the beam steering in a three-dimensional space around the end-fire direction of the antenna can be performed by activating a half cylinder of CFL array, while the height and radius of the cup change the scanning angle. Although the dimensions of the commercially available plasma tubes used in this study enforce the height of the plasma cup, in practical applications customized plasma tubes can be used to bypass the barrier of steering angle.

In short, through full-wave electromagnetic simulations it is demonstrated that, using a dense array of plasma CFLs instead of a metallic cup, the radiated beamwidth of a helical antenna can be electrically controlled, while its direction has the potential to be steered in the space.

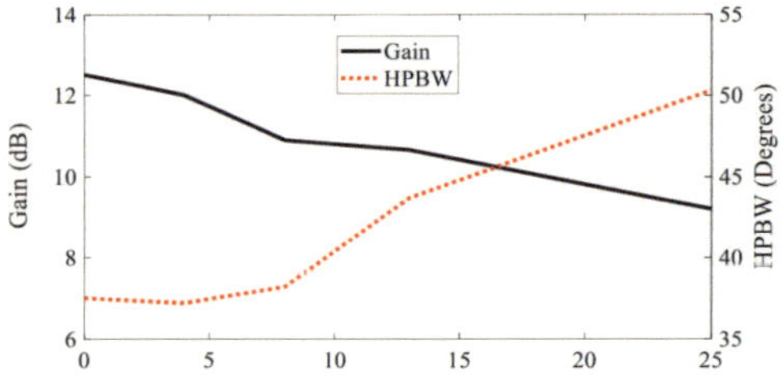

Figure 8.
Variations of the simulated radiation gain and HPBW of the cupped ground helical antenna versus the number of symmetrically excited plasma reflectors.

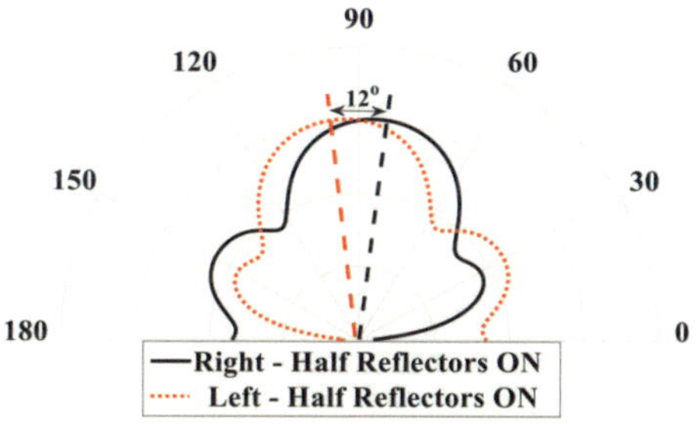

Figure 9.
Simulation results of the steered radiation pattern of the helical antenna using activation of the half number of CFLs.

3.1.3 Realization of the antenna and measurement results

The details of the realization and experimental validation of the plasma-cupped-ground helical antenna are devoted in this section. To this end, a prototype of the antenna is implemented and a comparison between simulated and measured results is presented.

By assembling the main body of the helical antenna and the plasma CFLs on a circular ground structure, as shown in **Figure 10a** and **b**, a prototype of the proposed antenna is realized. In this structure, each CFL is supplied by an electrical ballast which is connected to a relay. The states of each relay are controlled independently by a microcontroller board and as a result, every single CFL can be switched ON or OFF independently to control the radiation beamwidth or to steer the radiated beam toward the desired direction. The block diagram of the details of the reflector structure, including the CFLs and controlling electronic devices, is shown in **Figure 10c**.

Measurements are performed in an anechoic chamber with peak gain accuracy better than 0.5 dBi within the operating frequency range of the antenna. A comparison of simulated and measured antenna radiation gains for various plasma reflector configurations is shown in **Figure 11**. The measured gain and HPBW in the fully off state are 12.2 dBi and 42o, respectively, slightly wider than the simulated HPBW in this state. The measured gains for the all-element active and 13 alternately active configurations are 9.18 and 10.07 dBi, respectively, and the HPBW are 51.5o and 42.4o, respectively. In short, the good agreement between the simulation and

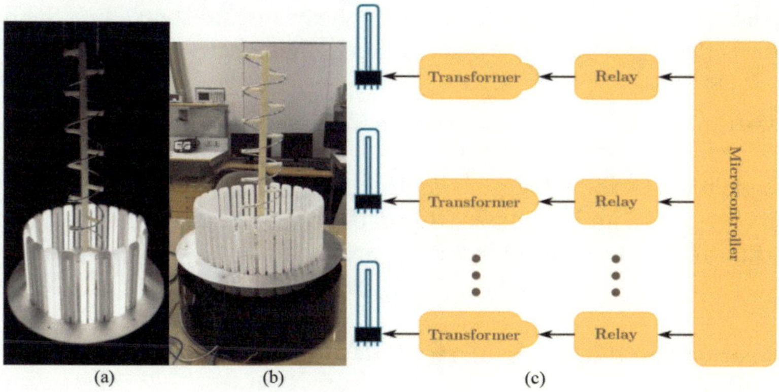

Figure 10.
(a), (b) Fabricated antenna, (c) Block diagram of the details of the structure of the showing antenna.

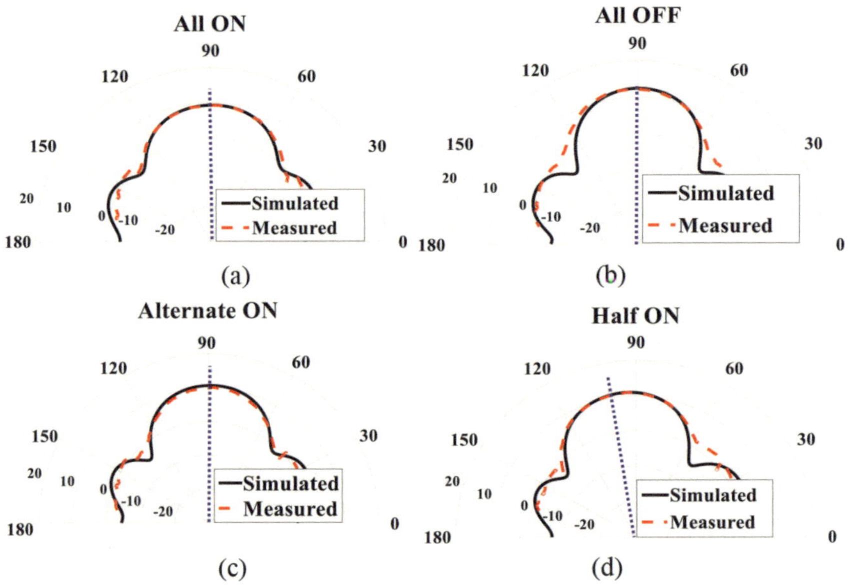

Figure 11.
Comparison of simulated and measured radiation gains of a plasma dish antenna at a frequency of 927 MHz for various configurations: (a) all CFLs active; (b) all CFLs are deactivated; (c) CFL is activated alternately; and (d) half of the adjacent CFL becomes active.

measurement results of the configuration confirms the simulation results related to the proposed method of electronic control of the antenna beamwidth. In total, 25 CFLs (odd number of CFLs) are used in this study to fully cover the cup bottom situation. Therefore, this mode is not a completely symmetrical configuration because alternating CFL activations leave two adjacent CFLs both activated (or deactivated). Therefore, some asymmetry in the radiation gain is expected. The time required to change the width or direction of the radiation beam depends only on the plasma decay time. This is about a few microseconds.

To experimentally verify the beam-steering capability of the proposed antenna, only 12 CFLs on one side of the bottom of the plasma cup are turned on and the remaining CFLs are turned off. The measured gain, in this case, is about 11.07 dBi at 96o, which matches very well with the simulated gain of 11.10 dBi, as shown in **Figure 11d**. In summary, the proposed technique for steering the beam and controlling the beamwidth using plasma reflectors is validated both numerically and experimentally.

3.1.4 Application of a half-cup plasma reflector for beam steering

An investigation on the effects of the dimensions of a half-cup plasma reflector, including its height HC and diameter DC, on the radiation characteristics of the helical antenna to control its steering angle will be presented in this section [46].

4. The effect of the diameter of the half-cup plasma reflector

To investigate the effect of the diameter of the half-cup D_C on the radiation characteristics of the antenna, a parametric study is conducted in this section. In this

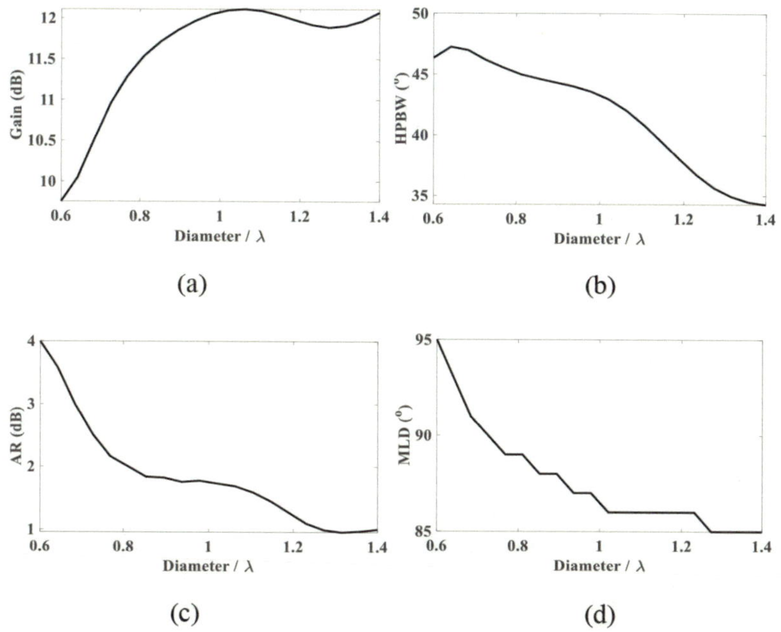

Figure 12.
Radiation characteristics of the plasma half-cupped-helical antenna with a diameter of the cup at H_C = 0.5 λ: (a) gain, (b) half-power beamwidth (HPBW), (c) axial ratio (AR), and (d) main lobe direction (MLD).

study, the diameter of the half-cup is changed between D_C = 0.6 λ and D_C =1.4 λ, while its height is fixed at H_C = 0.5 λ. **Figure 12** shows the simulated radiation characteristics of the antenna for different values of D_C. As shown in this figure, increasing the diameter of the half-cup reflector increases the gain while decreases the HPBW and AR of the antenna. Moreover, variations of D_C also change the direction of the radiated beam. The figure shows that for small diameters of the half-cup reflector, e.g., D_C < 0.7 in **Figure 12c**, the polarization of the antenna is not circular because the AR is greater than 3dB. Furthermore, as shown in **Figure 12d**, for small values of D_C, the direction of the main beam is greater than 90°, i.e., the radiated beam is tilted toward the plasma half-cup. In contrast, further increasing the diameter of the half-

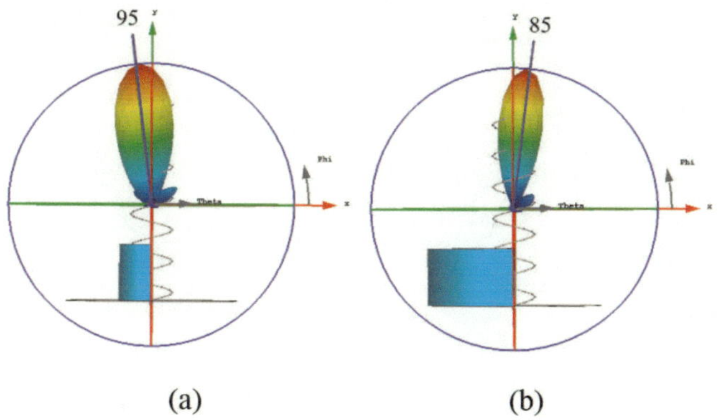

Figure 13.
Simulated radiation pattern of the plasma half-cupped-helical antenna at H_C = 0.5 λ when: (a) D_C = 0.6 λ, (b) D_C = 1.4 λ.

cup to the greater diameters than 0.7 λ, changes the direction of the main beam to the opposite side with respect to the half-cup, i.e., to angles smaller than 90 degrees. Based on these graphs, an optimum diameter of around 1.0 λ for the half-cup is also observed to achieve the maximum gain of 12.2 dBi with the circular polarization, while the beam is steered to 85°. In short, altering the diameter of the half-cup controls the direction of the main beam of the helical antenna, as shown in 3D radiation plots in **Figure 13** for two different diameters of the half-cup plasma reflector at H_C = 0.5 λ.

5. The effect of the height of the half-cup

Based on the results presented in the previous section, to compromise between a higher radiation gain and a lower AR for the antenna, a diameter of D_C = 1.0 λ is considered as the diameter of the half-cup in this section. By fixing D_C, the height of the half-cup plasma reflector (H_C) controls the physical shape of the half-cup. Thus, a parametric study is presented in this section based on the variation of H_C. The numerical results of the radiation characteristics of the antenna versus H_C, when it changes between H_C = 0 and H_C = 1.5 λ, are shown in **Figure 14**. The graphs clearly show that H_C changes all the radiation characteristics of the antenna. Specifically, the figure shows that for heights of smaller than 0.7 λ (H_C < 0.7 λ) the polarization of the antenna is circular (AR < 3 dB), and the maximum gain is achieved when $H_C \approx 0$, i.e. when there is no conductor reflector around the helix. The figures also show that increasing H_C from H_C = 0 to H_C = 0.7 λ decreases the radiation gain while steers the beam of the antenna from end-fire (i.e., θ = 90°) to θ = 84° in the opposite direction of the half-cup. Further increasing the H_C from 0.7 λ, increases the steering angle of the

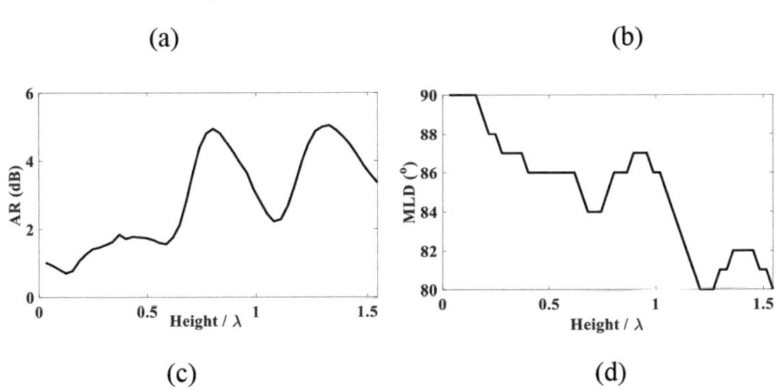

Figure 14.
Simulated radiation characteristics of the half-cupped-plasma antenna with a height of the cup at D_C = 1.0 λ: (a) gain, (b) HPBW, (c) AR, and (d) MLD.

DOI: http://dx.doi.org/10.5772/intechopen.108017

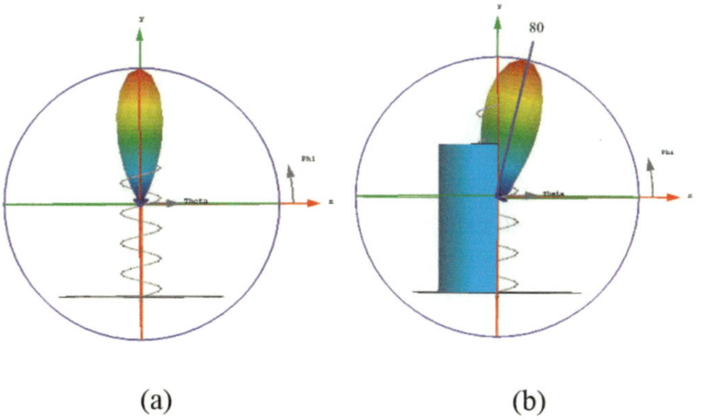

Figure 15.
Radiation of the half-cup plasma antenna with D_C = 1.0 λ when: (a) H_C = 0.0 λ, (b) H_C =1.25 λ.

main beam up to θ = 80°. However, the polarization of the antenna is not circular in this case. To have a better insight, 3-D radiation plots of the antenna for two special cases of H_C = 0.0 λ and H_C = 1.5 λ when D_C = 1.0 λ are shown in **Figure 15**.

In summary, from these graphs, it can be concluded that increasing the height of the half-cup steers the radiated beam of the helical antenna which is the target of this study. However, an appropriate selection of the height of the cup is required to achieve the maximum steering angle while maintaining the circular polarization using a half-cup plasma reflector around the helical antenna.

5.1 Summary

Dynamic beamwidth control and beam steering using plasma technology for the radiation beam of axial mode helical antennas is proposed. It has been shown that a circular array of plasma CFLs around the helical antenna controls electronically the direction and width of the radiated beam of the antenna. Fabrication of prototypes of the proposed structure and measurements of the radiation properties of various configurations of plasma reflectors confirmed the concept and computational results. The proposed antenna can be considered as a cost-effective solution for scanning and finding the target in certain applications related to space communications and radar systems.

5.2 Plasma helicone antenna

As previously mentioned in Section 3.1, an efficient technique to improve the radiation characteristics of a conventional axial mode helical antenna is shaping the ground conductor [40–44]. Some papers proposed a truncated conical reflector around an axial-mode helical antenna, known as a helicone antenna, in order to enhance its radiation gain, up to 5 dB, and also to minimize the side-lobe levels of helical antennas [42, 47, 48]. To compromise between the maximum radiation gain and minimum physical dimensions of the antenna, the optimal dimensions of the conical ground plane of the antenna have been presented in [42]. This antenna has applications in many communication systems where high radiation gain is desirable, including tracking and space communication systems.

Generally, a helicone antenna covers an area with a certain footprint. When the location of the target is known, a high-gain antenna will be preferred and this approach may be effective. However, when the location of the target is inaccurate, misalignment loss, especially in space applications, may be very high. In this case, if the beamwidth of the antenna in the ground station can be varied, even with less radiation gain, an alternate approach may be used. Using auto-tracking antenna pedestals is one method to address this issue. However, controlling the beamwidth of the ground station antenna is an alternative elegant method so that in normal conditions the antenna is adjusted to have its maximum gain, and when the target is lost, to find its accurate position its beamwidth is increased. For instance, to cover a relatively wide area, a conventional helical antenna with a relatively wide beamwidth can be used, and after receiving some information from the target, it can be reconfigured to have a narrower beamwidth (and as a result higher gain). The required reconfiguration from an antenna with a wider HPBW to an antenna with a higher gain is possible if the truncated conical ground of the antenna can be electrically switched ON and OFF or its dimension can be altered.

Although there are different techniques for controlling the beamwidth of a single antenna [49–52], it is critical that the selected technique minimizes the complexity of the hardware and software designs and as a result and reduces the costs associated with such an antenna. The aim of this study is to demonstrate a truncated conical plasma reflector around a conventional axial mode helical antenna as an efficient and relatively low-cost method to enhance the radiation gain and control HPBW of a helical antenna. It will be shown that there will be two radiation modes for the proposed antenna, namely, a narrow beam mode with higher radiation gain and a wide beam mode with lower radiation gain. To verify the performance of the proposed antenna, the simulation results of the designed antenna in terms of radiation gain and HPBW are presented. The results show that there are two different operating modes based on the state of the plasma. In the first mode, the plasma reflectors are switched OFF and the proposed structure acts as a conventional helical antenna, and in the second mode, the plasma reflectors are switched ON and the structure acts as a helical antenna with a truncated conical plasma reflector. As expected, in the second mode, the radiation gain is higher than the gain in the first mode. For the realization of the proposed structure, it is important to consider the feasibility, level of complexity, and the costs associated with the implementation. Thus, based on these considerations, different techniques for the implementation are analyzed and the best structure in terms of ease of fabrication is selected.

5.2.1 The basic structure of a helicone antenna

Figure 16 illustrates the structure of the helicone antenna which is composed of a conventional axial-mode helical antenna with a truncated conical ground. In this figure, D, C, α, N, and L are, respectively, the diameter, the circumference, the pitch angle, the number of turns, and the length of the helical section of the antenna. For the truncated conical section, D_1, D_2, β, and H are, respectively, the inner diameter, the outer diameter, the angle, and the height. The operating frequency of the antenna is 1280 MHz. At this frequency, the dimensions of the helical section of the antenna are D = 8.4 cm, C = 1.12 λ, α =11.35°, N = 5, and $L \approx$ 26.49 cm. Considering a circular ground plane with the radius of D_g = 1.5 D for the designed helical antenna, which is used as a reference in this study, the simulated radiation gain and beamwidth

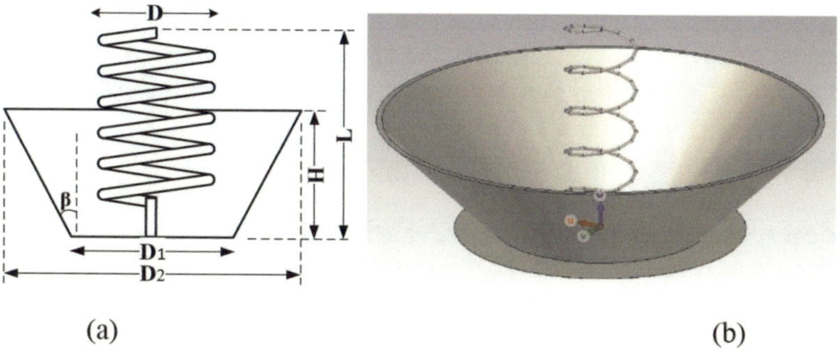

Figure 16.
(a) The structure of a conventional helicone antenna, (b) an illustration of the antenna.

are, respectively, G = 12.6 dBi and HPBW = 40.6° while the direction of the main beam of the helix is in the end-fire direction at θ= 90°.

Based on the study presented in [42], the optimum dimensions of the truncated conical ground section for the maximum gain are D_1 = 1.1 λ, H = 0.6 λ, and β =45°, which for the operating frequency of 1280 MHz, D_1 = 25.78 cm, and H = 14.06 cm. For the optimized helicone antenna, the simulated radiation gain and beamwidth are, respectively, G = 15 dBi and HPBW = 31.7° in the end-fire direction, while the polarization of the antenna is still circular. So, around 19% enhancement in the radiation gain of the optimized helicone antenna with respect to a circular ground plane helical antenna is observed while, as expected, the HPBW of the antenna has proportionally decreased.

5.2.2 A helicone antenna using plasma conical ground

In this section, a helicone antenna using a truncated conical ground of the plasma material is analyzed numerically. Here, the plasma and collision frequencies are, respectively, ω_p = 14 $\times 10^{10}$ rad/s and ν = 400 MHz, the height of the cone is 0.6 λ and its angle is 45°. The simulation results show a gain of G = 15.3 dBi and beamwidth of HPBW = 31.2° for the antenna, which proves that the performance of the plasma conical ground is similar to the performance of the metallic conical ground. Moreover, it is clear that by switching OFF the plasma section, the radiation gain and beamwidth of the antenna, just like a single helical antenna, are, respectively, G = 12.6 dBi and HPBW = 40.6°, as shown in **Figure 17**. So, the helical antenna with plasma conical ground is capable of switching between two different radiation patterns, by switching the plasma conical ground ON and OFF. When the plasma conical ground is switched

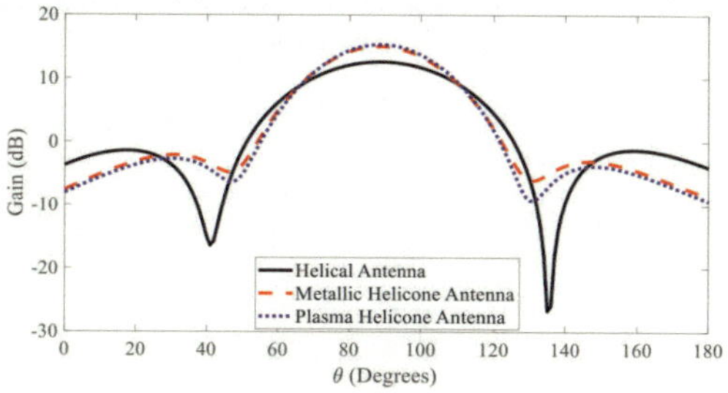

Figure 17.
The simulated radiation gain of the helicone antenna using a plasma conical reflector.

ON, a helicone antenna is achieved which benefits from a higher radiation gain (or narrower beamwidth) and when the plasma reflector is switched OFF a helical antenna is achieved with lower radiation gain (or wider beamwidth).

5.2.3 The implementation of the plasma helicone antenna

Numerical results presented in the previous section show that a truncated conical plasma reflector provides a switchable beamwidth (or gain) for the helical antenna. In practice, the realization of the homogenous conical section by plasma material is not straightforward. So, the aim of this section is to analyze different conical structures based on the commercial fluorescent lamps for the realization of the plasma conical reflector. To this end, different structures are analyzed and their performances are evaluated.

5.2.4 Plasma helicone antenna based on the U-shaped compact fluorescent lamps

As the first method for the realization of the plasma helicone antenna, as shown in **Figure 18**, a sufficiently dense array of U-shaped CFLs are arranged in a conical structure. The dimensions of a U-shaped CFL has been shown in **Figure 6b**. The number of CFLs in the structure is selected based on the circumference of the bottom face of the cone so that a dense array of CFLs covers this face. The length of each CFL is 18 cm from the circular ground and the diameter of each CFL column is 14.7 mm, while a 1.6 mm space gap is between the columns. So, to form the truncated conical reflector, 19 CFLs are arranged at the bottom face of the cone. The angle between the ground plane and each CFL is 45°. Note that the packaging of the commercial off-the-shelf CFLs is such that small space gaps between adjacent elements are inevitable. It is clearly observed in the figure that the distance between the CFLs in the larger diameter of the cone (i.e. D_2) is not possible to be ignored. So, discrepancies between the performance of the conical-shaped homogenous plasma reflector and the conical reflector using U-shaped CFLs are expected. The simulated radiation gain and beamwidth of the plasma helicone antenna with based on the proposed structure are G = 12.8 dBi and HPBW = 37.8° compared to G =12.6 dBi and HPBW = 30.6° for the helical reference antenna (without the conical ground). This structure shows less than 2% improvement in the radiation gain of the antenna, while the total power consumption of the antenna is 342 W.

5.2.5 Plasma helicone antenna based on the fluorescent tubes

In this case, 45 commercial conventional fluorescent lamps are arranged in a conical shape to form the reflector, as shown in **Figure 19**. The diameter and length of

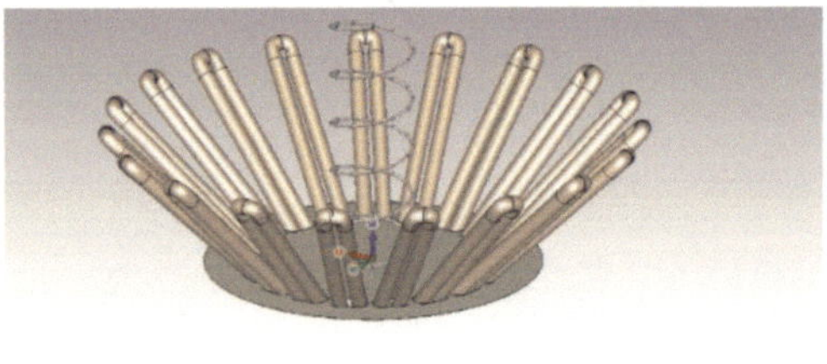

Figure 18.
The structure of the helicone antenna using U-Shaped CFLs.

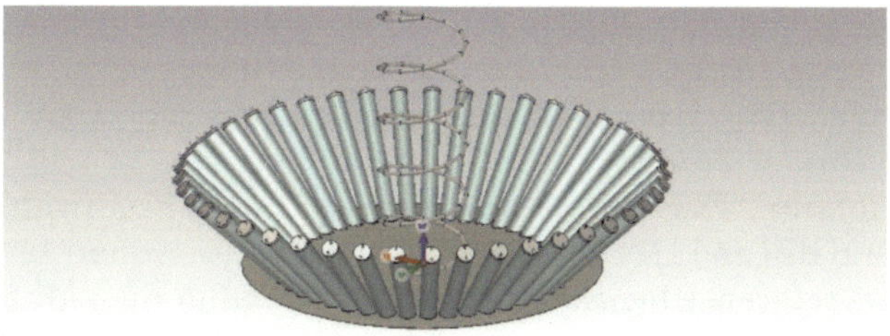

Figure 19.
The structure of the helicone antenna using commercial fluorescent lamps.

the fluorescent lamps are 1.5 cm and 18 cm, respectively. The plasma and collision frequencies are, respectively, $\omega_p = 14 \times 10^{10}$ rad/s and $\nu = 400$ MHz for each lamp. Since the power consumption of each lamp is 6 W, the total power consumption of this structure is 270 W. The simulated radiation gain and HPBW of the helicone antenna based on this structure are, respectively, G = 13.1 dBi and HPBW = 37° when all the lamps are excited. This shows a 3% improvement in the radiation gain with respect to the reference helical antenna. It seems that this structure is also inefficient because the distance between the plasma elements in the upper section of the conical reflector is large.

5.2.6 Plasma helicone antenna based on the combination of U-shaped CFLs and fluorescent lamps

In order to improve the radiation gain of the plasma helicone antenna, a new version of the truncated conical reflector is proposed here based on the combination of the U-shaped CFLs and cylindrical fluorescent lamps. In this structure, 19 U-shaped CFLs are arranged in the structure shown in **Figure 18**, while 19 fluorescent lamps are also inserted between them from the upper side of the cone, as shown in **Figure 20**. The holder of the fluorescent tubes is made from Teflon. The simulated radiation gain and beamwidth of this structure are, respectively, G = 14.7 dBi and HPBW = 30°, which around 17% enhancement in the gain is observed with respect to the reference helical antenna. The power consumption of this structure is around 459 W.

5.2.7 Plasma helicone antenna based on a spiral plasma tube

As another alternative for the implementation of the conical ground in the helicone antenna, the effects of a spiral plasma lamp are also investigated. A view of the

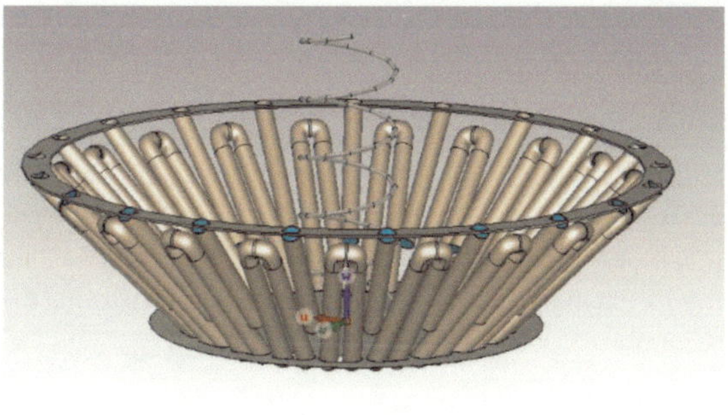

Figure 20.
3-D view of a helicone antenna based on the combination of fluorescent lamps and U-shaped CFLs.

proposed structure is shown in **Figure 21**. The diameter and thickness of the spiral plasma tube are, respectively, 1 cm 0.1 cm while the tube material is Pyrex. To form the conical plasma section, around 17 turns of the plasma tube are wrapped in this study. The plasma and collision frequencies are around $\omega_p = 14 \times 10^{10}$ rad/s and $\nu =$ 400 MHz in the column. The simulated radiation gain and beamwidth of this antenna are around G = 14.6 dBi and HPBW = 33.9°, respectively. Around 16% enhancement with respect to the reference helical antenna is observed in the radiation gain using this structure which is similar to the enhancement of the radiation gain in the previous case in which a combination of fluorescent lamps and U-shaped CFLs are used for the implementation of the conical section. A comparison between the simulated gain of the helical antenna using proposed plasma reflectors in the two last cases with the gain of the reference helical and helicone antennas is shown in **Figure 22**.

5.2.8 Reconfigurable slotted cylindrical waveguide and coaxial array plasma antenna

Slotted waveguide antennas have been studied in the literature for many years [53–59]. A number of papers [60–64] have proposed plasma-filled waveguides for mode modification or as protection against high-power electromagnetic waves. In this study, the waveguide is partially filled with plasma. In this section, we propose a reconfigurable cylindrical slotted waveguide antenna coupled to a plasma tube, and study the radiation pattern, gain, S-parameters, and we present simulation and experimental results related to efficiency and performance. This antenna system offers him two modes of operation depending on the plasma conditions. 1) the waveguide mode, which is active when the plasma is off, and 2) the coaxial mode, when the plasma is on. Performance of the reconfigurable system is observed in terms of reflection coefficient, maximum realizable gain, radiation pattern, and overall efficiency. Slotted waveguide antennas have been studied in the literature for many years

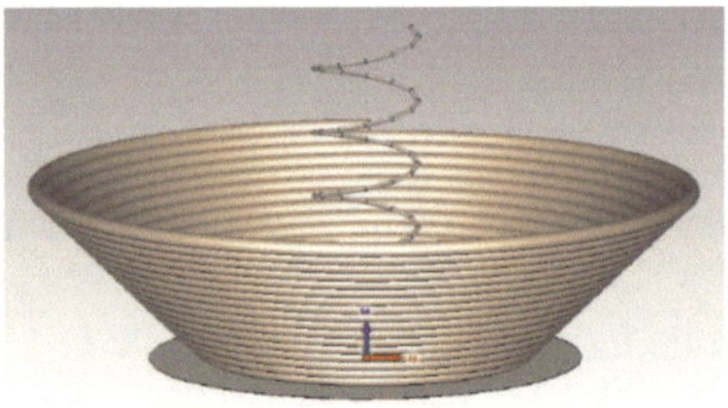

Figure 21.
The structure of the helicone antenna using conical spiral plasma lighting tubes.

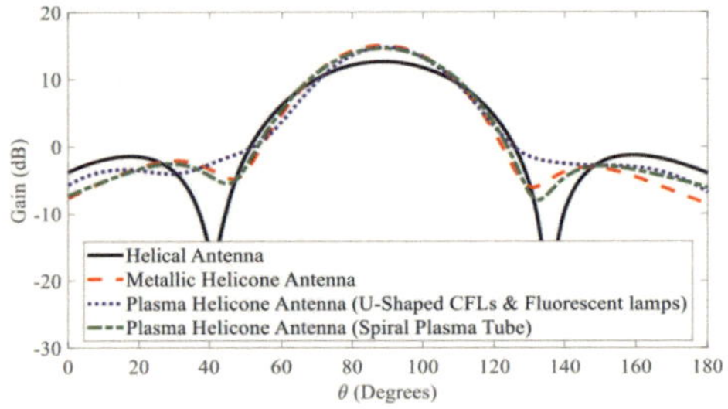

Figure 22.
The simulated radiation gain of the helicone antenna for different cases.

[53–59]. A number of papers [60–64] have proposed plasma-filled waveguides for mode modification or as protection against high-power electromagnetic waves. In this study, the waveguide is partially filled with plasma. In this section, we propose a reconfigurable cylindrical slotted waveguide antenna coupled to a plasma tube, and study the radiation pattern, gain, S-parameters, and We present simulation and experimental results related to efficiency and performance. This antenna system offers her two modes of operation depending on the plasma conditions. 1) the waveguide mode, which is active when the plasma is off, and 2) the coaxial mode, when the plasma is on. Performance of the reconfigurable system is observed in terms of reflection coefficient, maximum realizable gain, radiation pattern, and overall efficiency.

5.3 Modeling and simulation

The shape of the waveguide is shown in **Figure 23**. The diameter and length of the metal tube are 70 mm and 620 mm, respectively. A plasma tube, which is a commercially available fluorescent lamp, is inserted into the metal tube (see **Figure 24a** and **b**). The height and diameter of the lamp are 590 mm and 26 mm, respectively. The feed system consists of a metal ring surrounding both ends of the lamp and a coaxial cable (see **Figure 24a**) attached to this metal ring to provide the RF signal. The realized prototype is shown in **Figure 24b**. In this study, CST Microwave Studio is used for simulation. The gas-containing tube consists of glossy glass with εr = 4.82, tan δ = 0.005 and a thickness of 0.5 mm. The plasma follows the Drude model [65, 66]

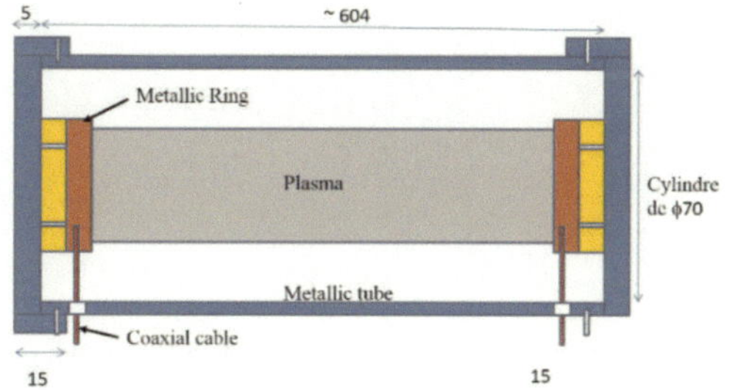

Figure 23.
Antenna shape.

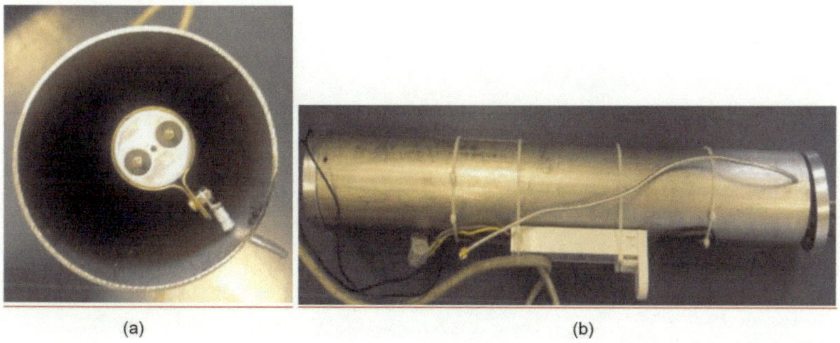

Figure 24.
Realized models. (a) Feeding lamp. (b) Realized waveguide.

defined by two parameters: plasma angular frequency $\omega p = 43.9823 \times 109$ rad/s and electron neutral collision frequency $\nu = 900$ MHz.

5.4 Results and discussion

5.4.1 Plasma waveguide antenna

First, we investigate the S-parameters of the antenna and examine the behavior of the metal tube depending on the plasma state. **Figure 25** shows simulated S11 and S21 of the proposed structure for both plasma off and plasma on cases. Note that when the plasma is turned off, we expect a conventional waveguide with a cutoff frequency of about 2.5 GHz, as shown in **Figure 25a**. However, when the plasma is on, the coaxial conductor appears with stronger absorption around 4 GHz, as shown in **Figure 25b**.

In order to validate the simulation results, S21 measurements were carried out in cases of plasma activation and deactivation, as shown in **Figure 26**. Simulations and measurements agree, but the observed loss is more significant. Plasma OFF curves indicate low-frequency lamps, possibly due to prototype and delivery system manufacturing.

5.4.2 Slotted waveguide antenna

Now, to use the metal tube as an antenna, make 12 small slits (ls = 40mm, ws = 5mm) in the surface of the tube, as shown in **Figure 27**. The distance between the slots

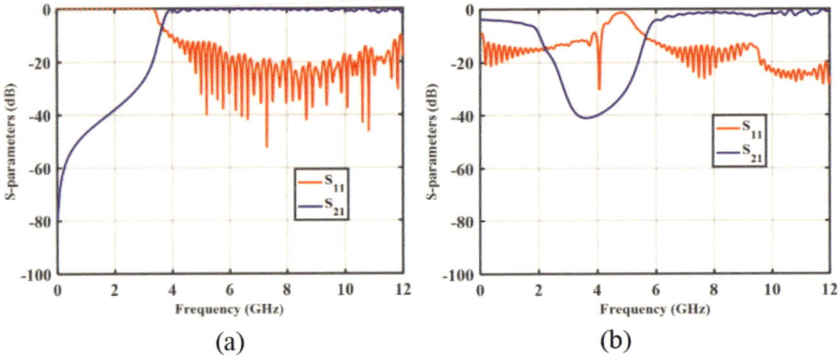

Figure 25.
Simulated S11 and S21 magnitude. (a) Plasma is OFF. (b) Plasma is ON.

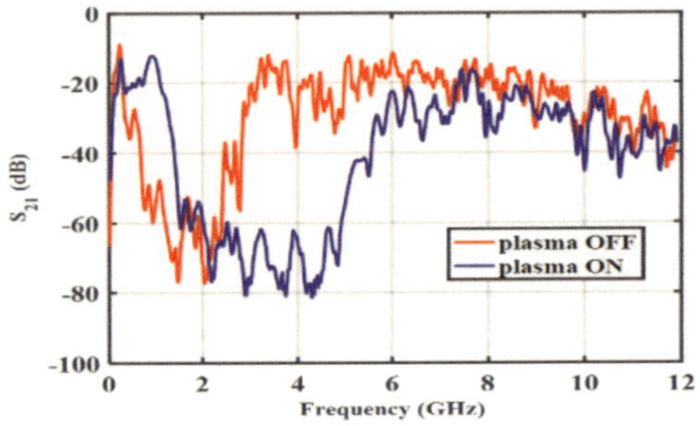

Figure 26.
Measured S21 magnitude when the plasma is OFF and ON.

Figure 27.
Slotted waveguide geometry, small slot for 4 GHz antenna (ls = 40 mm, ws = 5 mm) and large slot for 1 GHz antenna (ls = 150 mm, ws = 5 mm).

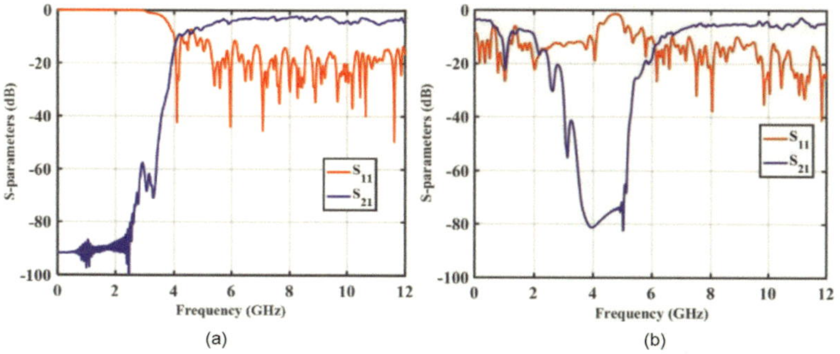

Figure 28.
Simulated sizes of S11 and S21 for slot antennas. (a) Plasma is off. (b) Plasma is on.

is λ/2 to 4 GHz (40 mm), so the antenna radiates up to 4 GHz with the plasma turned off. It also has four large slots (ls = 150 mm, ws = 5 mm) and a slot spacing of λ/2 at 1 GHz (150 mm). The antenna operates at 1 GHz when the plasma is on.

Figure 28 shows the S11 and S21 simulations of the slotted antenna both when the plasma is off (**Figure 28a**) and on (**Figure 28b**). The match around 4GHz is -40dB and the S21 is -10dB. This means that most of the power is radiated resulting in good gain and efficiency. By turning on the plasma, the 1 GHz match due to the waveguide to coaxial line conversion is -20 dB and S21 is about -20 dB. In this case, a 4-element array provides good gain and efficiency.

A radiation pattern is shown in **Figure 2** to demonstrate the reconfigurable capability of this antenna system. **Figure 29a** and **29b** show simulated E-plane radiation patterns for 4 GHz frequency when plasma is off and 1 GHz frequency when plasma is on, respectively. In the two simulation results, each radiation pattern is normalized to

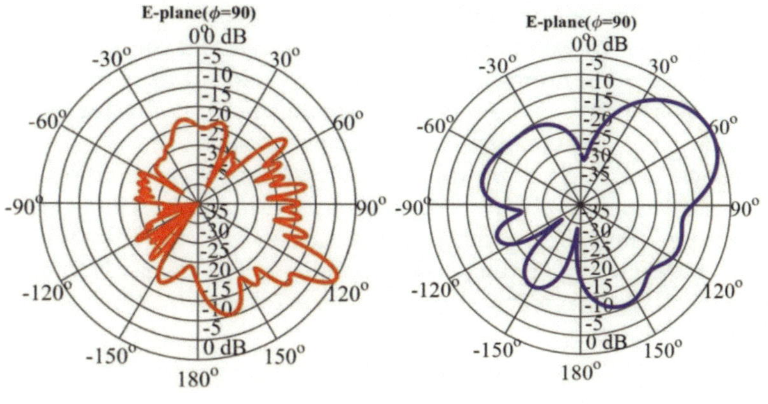

Figure 29.
Normalized H-plane radiation pattern. (a) 4 GHz with the plasma turned off. (b) 1 GHz with plasma ON.

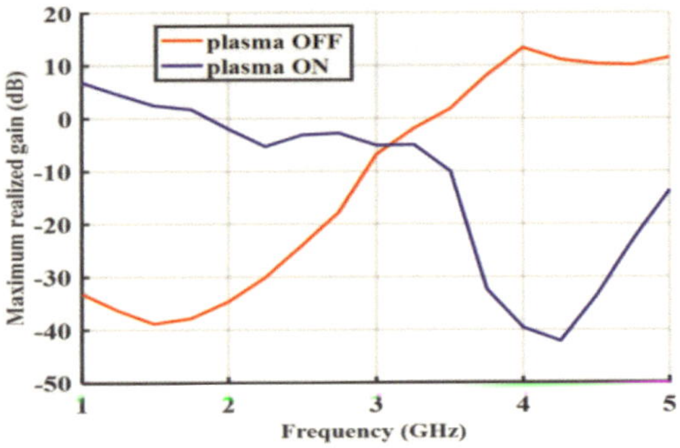

Figure 30.
Gain of the antenna in plasma OFF and ON cases.

its electric field maximum. The -3 dB simulation bandwidths are 8.1° to 4 GHz and 34.1° to 1 GHz. The antenna is tilted between two frequencies, 120° to 4 GHz and 60° to 1 GHz. Sidelobe level (SLL) is close to -10dB regardless of configuration.

Figure 30 shows the maximum simulation gain achieved. For plasma shutdown, the gain is maximum at 4 GHz, 13.27 dBi. In the case of ON, the gain is maximal at 1 GHz and is equivalent to 6.82 dBi. As far as gain is concerned, there is a 40 dB to 1 GHz difference between plasma OFF and plasma ON. The difference of 52 dB to 4 GHz between switched-off plasma and activated plasma. Normally, the antenna is not expected to operate at 1 GHz because the cut-off frequency is greater than 1 GHz. But once the lamp is switched on, the waveguide becomes a coaxial line from which to obtain radiation and gain at 1 GHz. Gain allows evaluation of antenna efficiency. Total efficiency is 83.4% to 4 GHz (extinguished plasma) and 68.2% to 1 GHz (activated plasma).

6. Conclusion

In this chapter, several approaches for the dynamic control of the beamwidth and radiation gain of circular polarized helical antennas based on plasma reflectors have been presented. The concept and design principles were discussed and confirmed through complete full-wave simulations. Various achievements of the plasma reflector, using U-shaped CFLs, commercial fluorescent tubes and conical spiral plasma tubes have been explored. From the upper and lower side of the cone, the most effective solution is to replace the metallic truncated conical reflector by a combination of U-shaped CFLs and commercial fluorescent tubes.

In all cases, we look at the practicalities and costs of implementation.

In the last section, a reconfigurable slot antenna using a plasma tube inside a metal waveguide was presented. The antenna exhibits reconfigurable radiation patterns at 4 GHz and 1 GHz, depending on the plasma conditions. The main advantage of this antenna with the plasma on is that it can radiate at 1 GHz while the gain drops off sharply at 4 GHz frequency.

Author details

Fatemeh Sadeghikia[1]*, Ali Karami Horestani[1] and Mohamed Himdi[2]

1 Wireless Telecommunication Group, ARI, Ministry of Science, Research and Technology, Tehran, Iran

2 IETR, University of Rennes 1, Rennes, France

*Address all correspondence to: sadeghi_kia@ari.ac.ir

References

[1] Sadeghikia F, Dorbin MR, Horestani AK, Noghani MT, Ja'afar H. Tunable inverted F antenna using plasma technologies. IEEE Antennas and Wireless Propagation Letters. 2019; **18**(4):702-706

[2] Ja'afar H, Ali MTB, Dagang ANB, Zali HM, Halili NA. A reconfigurable monopole antenna with fluorescent tubes using plasma windowing concepts for 4.9-GHz application. IEEE Transaction on Plasma Science. 2015;**43**(3):815-820

[3] Alexeff I, Anderson T, Parameswaran S, Pradeep EP, Hulloli J, Hulloli P. Experimental and theoretical results with plasma antennas. IEEE Transactions on Plasma Science. 2006; **34**(2):166-172

[4] Jusoh MT, Lafond O, Colombel F, Himdi M. Performance and radiation patterns of a reconfigurable plasma corner-reflector antenna. IEEE Antennas and Wireless Propagation Letters. 2013; **12**:1137-1140

[5] Yuan X, Li Z, Rodrigo D, Mopidevi H, Kaynar O, Jofre L, et al. A parasitic layer-based reconfigurable antenna design by multi-objective optimization. IEEE Transactions on Antennas and Propagation. 2012;**60**(6):2690-2701

[6] Plapous C, Cheng J, Taillefer E, Hirata A, Ohira T. Reactance domain music algorithm for electronically steerable parasitic array radiator. IEEE Transactions on Antennas and Propagation. 2004;**52**(12):3257-3264

[7] Fourikis N. Phased Array-Based Systems and Applications. Newyork: Wiley; 1997

[8] Jusoh MT, Lafond O, Colombel F, Himdi M. Performance of a reconfigurable reflector antenna with scanning capability using low-cost plasma elements. Microwave and Optical Technology Letters. 2013;**55**(12): 2869-2874

[9] Barro OA, Lafond O, Himdi H. Reconfigurable antennas radiations using plasma Faraday cage. In: 2015 international Conference on Electromagnetics in Advanced Applications (ICEAA). Turin, Italy: ICEAA; 07-11 Sep 2015. pp. 545-548

[10] Sadeghikia F, Noghani MT, Simard M. Experimental study on the surface wave driven plasma antenna. AEU International Journal of Electronic Communication. 2016;**70**:652-656

[11] Sadeghikia F, Dorbin MR, Horestani AK, Noghani MT, Ja'afar H. Multi-beam frequency tunable antenna based on plasma-nested helix. In: 2019 13th European conference on antennas and propagation (EuCAP). Krakow, Poland: EuCAP; 31 Mar - 05 Apr 2019. pp. 1-3

[12] Dorbin MR, Mohassel R, Sadeghikia F, Binti Ja'afar H. Analytical estimation of the efficiency of surface-wave-excited plasma monopole antennas. IEEE Transactions on Antennas and Propagation. 2022;**70**(4): 3040-3045

[13] Dorbin MR, Rashed Mohassel JA, Sadeghikia F, Ja'afar HB. Analytical Estimation of the Efficiency of Surface-Wave-Excited Plasma Monopole Antennas. In: IEEE Transactions on Antennas and Propagation. Vol. 70. No. 4. Apr 2022. pp. 3040-33045

[14] Sadeghikia F, Doorbin MR, Jaafar H, Horestani AK, Noghani MT. An overview on the implementation of

surface wave driven plasma antennas. In: IEEE Symposium on Wireless Technology & Applications (ISWTA). Malaysia; 2021. pp. 53-57

[15] Sadeghikia F, Horestani AK, Dorbin MR, Noghani MT, Ja'afar H. A new feed network for the communication signal and excitation of surface-wave-driven plasma antennas. In: 2020 14th European Conference on Antennas and Propagation (EuCAP). Copenhagen, Denmark: EuCAP; 15-20 Mar 2020. pp. 1-4

[16] Noghani MT, Horestani AK, Sadeghikia F, Dorbin MR. Theoretical modeling of resonant wavelength in 3-layered plasma antennas. Waves in Random and Complex Media. 2021; **31**(6):1587-1596

[17] Sadeghikia F, Hodjat-Kashani F, Rashed-Mohassel J, Ghayoomeh-Bozorgi J. The effects of the tube characteristics on the performance of a plasma monopole antenna. In: Proc. 32nd PIERS Conf. in Moscow. Russia; 2012. pp. 1208-1211

[18] Sadeghi-Kia F. Analysis of plasma monopole antenna using numerical method and an equivalent circuit. IEEE Antennas and Wireless Propagation Letters. 2017;**16**:1711-1714

[19] Sadeghikia F, Hodjat-Kashani F. A two element plasma antenna array. Engineering, Technology & Applied Science Research. 2013;**3**(5):516-521

[20] Sadeghikia F, Hodjat-Kashani F, Rashed-Mohassel J, Ghayoomeh-Bozorgi J. Characterization of a surface wave driven plasma monopole antenna. Journal of Electromagnetic Waves and Applications. 2012;**26**(2/3):239-250

[21] Sadeghikia F, Hodjat-Kashani F, Rashed-Mohassel J, Ghayoomeh-Bozorgi J. Characteristics of plasma antennas under radial and axial density variations. In: PIERS Conference in Moscow. Russia; 2012. pp. 1212-1215

[22] Sadeghikia F, Hodjat-Kashani F, Rashed-Mohassel J, Lotfi AA, Ghayoomeh-Bozorgi J. A Yagi-UDA plasma monopole array. Journal of Electromagnetic Waves and Application. 2012;**26**:885-894

[23] Sadeghikia F, Horestani AK, Noghani MT, Dorbin MR, Mahdikia H, Ja'afar H. A study on the effect of the radius of a cylindrical plasma antenna on its radiation efficiency. International Journal of Engineering and Technology. 2018;7:204-206

[24] Sadeghikiya F, Hojatkashani F, Rashed Mohasel J, Ghiyome Bozorgi SJ. Space application of a linear array of plasma antenna. Journal of Space Science and Technology. 2012;**5**(3):59-66

[25] Sadeghi-Kia F, Hodjat-Kashani F, Rashed-Mohassel J. Analysis of tapered column plasma using full wave simulator. In: Loughborough Antennas & Propagation Conference (LAPC). Loughborough, UK; 2009

[26] Sadeghikia F, Hodjat-Kashani F, Rashed-Mohassel J. An investigation on surface wave driven plasma monopole antenna. Journal of Iranian Association of Electrical and Electronics Engineers. 2013;**10**(1):7-16

[27] Noghani MT, Sadeghikia F, Dorbin MR, Horestan AK. Analysis of surface waves in a 3-layered cylindrical plasma antenna. In: 27th Iranian Conference on Electrical Engineering (ICEE). Iran; 2019. pp. 1425-1428

[28] Valipour M, Sadeghikia F, Karami-Horestani A, Ja'afar H. The effect of non-uniform conductivity on radiation

characteristics of a monopole plasma antenna. In: 13th European Conference on Antennas and Propagation (EUCAP). Krakow, Poland; 2019

[29] Horestani AK, Noghani MT, Sadeghikia F, Dorbin MR, Valipour M, Martín F. Reconfigurable and frequency tunable inverted F antenna based on plasma technology. In: International Conference on Electromagnetics in Advanced Applications (ICEAA). Granada, Spain; 2019. pp. 1175-1177

[30] Ja'afar H, Abdullah R, Redzwan FNM, Sadeghikia F. Analysis of cylindrical monopole plasma antenna design. In: International Symposium on Antennas and Propagation (ISAP). Busan, Korea (South); 2018. pp. 1-2

[31] Sadeghikia F, Noghani MT, Dorbin MR, Seyyedi J, Jabbari A. A study on the physical characteristics and development of plasma reflectors. Journal of Radar. 2021;7(1):35-44

[32] Sadeghikia F. Plasma antenna technology in space missions. Journal of Space Science and Technology. 2017; **10**(1):27-34

[33] Moisan M, Nowakowska H. Contribution of surface-wave (SW) sustained plasma columns to the modeling of RF and microwave discharges with new insight into some of their features. A survey of other types of SW discharges. Plasma Sources Science and Technology. 2018;**27**(7)

[34] Manheimer W. Plasma reflectors for electronic beam steering in radar systems. IEEE Transactions on Plasma Science. 1993;**19**(6):1228-1234

[35] Anderson T. Fundamental plasma antenna theory. In: Plasma Antennas. Norwood, MA, USA: Artech House; 2011

[36] Ha A, Chae MH, Kim K. Beamwidth control of an impulse radiating antenna using a liquid metal reflector. IEEE Antennas and Wireless Propagation Letters. 2019;**18**(4):571-575

[37] Chatterjee A, Parui SK. Beamwidth Control of Omnidirectional antenna using conformal frequency selective surface of different curvatures. IEEE Transactions on Antennas and Propagation. 2018;**66**(6):3225-3230

[38] Debogović T, Perruisseau-Carrier J. Array-fed partially reflective surface antenna with independent scanning and beamwidth dynamic control. IEEE Transactions on Antennas and Propagation. 2014;**62**(1):446-449

[39] Towfiq A, Bahceci I, Blanch SA, Romeu J, Jofre L, Cetiner BA. A reconfigurable antenna with beam steering and beamwidth variability for wireless communications. IEEE Transactions on Antennas and Propagation. 2018;**66**(10):5052-5063

[40] Sadeghikia F, Valipour M, Noghani MT, Ja'afar H, Horestani AK. 3D beam steering end-fire helical antenna with beamwidth control using plasma reflectors. IEEE Transactions on Antennas and Propagation. 2021;**69**(5): 2507-2512

[41] Djordjevic AR, Zajic AG, Ilic MM. Enhancing the gain of helical antennas by shaping the ground conductor. IEEE Antennas and Wireless Propagation Letters. 2006;**5**:138-140

[42] Sadeghikia F, Horestani AK. Design guidelines for helicone antennas with improved gain. Microwave Optics Technology Letters. 2019;**61**(4):1016-1021

[43] Sadeghkia F, Mahmoodi M, Hashemi-Mehneh H, Ghayoomeh J.

Helical antenna over different ground planes. In: 8th European Conference on Antennas and Propagation (EuCAP). The Hague, Netherlands: EuCAP; 06-11 Apr 2014. pp. 2185-2188

[44] Valipour M, Sadeghikia F, Karami-Horestani A, Himdi M. Beamwidth control of a helical antenna using truncated conical plasma reflectors. In: 14th European Conference on Antennas and Propagation (EuCAP). Denmark; 2020

[45] Sadeghikia F, Mahdikia H, Dorbin MR, Noghani MT, Horestani AK, Jaafar H. A study on the effect of gas pressure and excitation frequency of a cylindrical plasma antenna on its radiation efficiency. In: 13th European Conference on Antennas and Propagation (EuCAP). Poland; 2019

[46] Sadeghikia F, Valipour M, Horestani AK, Himdi M, Anderson T. Beam-steerable helical antenna using plasma reflectors. In: 16th European Conference on Antennas and Propagation (EuCAP). Madrid, Spain; 2022

[47] Olcan DI, Zajic AR, Ilic MM, Djordjevic AR. On the optimal dimensions of helical antenna with truncated cone reflector. In: 1st European Conference on Antennas and Propagation. Nice, France; 2006

[48] King H, Wong J. Characteristics of 1 to 8 wavelength uniform helical antennas. IEEE Transactions on Antennas and Propagation. 1980;**28**(2):291-296

[49] Weily AR, Bird TS, Guo YJ. A reconfigurable high-gain partially reflecting surface antenna. IEEE Transactions on Antennas and Propagation. 2008;**56**(11):3382-3390

[50] Deng WQ, Yang XS, Shen CS, Zhao J, Wang BZ. A dual-polarized pattern reconfigurable Yagi patch antenna for microbase stations. IEEE Transactions on Antennas and Propagation. 2017;**65**(10):5095-5102

[51] Debogovic T, Bartoli J, Perruisseau-Carrier J. Array-fed partially reflective surface antenna with dynamic beamwidth control and beamsteering. In: 2012 6th European Conference on Antennas and Propagation (EUCAP). Prague, Czech Republic: EUCAP; 26-30 Mar 2012. pp. 3599-3603

[52] Towfiq MA, Bahceci I, Blanch S, Romeu J, Jofre L, Cetiner BA. A reconfigurable antenna with beam rteering and beamwidth variability for wireless communications. IEEE Transactions on Antennas and Propagation. 2018;**66**(10):5052-5063

[53] Wang W, Jin J, Liang X-L, Zhang Z-H. Broadband dual polarized waveguide slotted antenna array. In: 2006 IEEE Antennas and Propagation Society International Symposium. Albuquerque, NM: IEEE; 09-14 Jul 2006. pp. 2237-2240

[54] Park S, Tsunemitsu Y, Hirokawa J, Ando M. Center feed single layer slotted waveguide array. IEEE Transactions on Antennas and Propagation. 2006;**54**(5): 1474-1480

[55] Harmouch A, Haddad HSA. Cylindrical omnidirectional slotted waveguide antenna with optimized directional characteristics. In: 2013 13th Mediterranean Microwave Symposium (MMS). Saida, Lebanon: MMS; 02- 05 Sep 2013. pp. 1-4

[56] Li T, Meng H, Dou W. Design and implementation of dual- frequency dual-polarization slotted waveguide antenna array for Ka- band application. IEEE Antennas and Wireless Propagation Letters. 2014;**13**:1317-1320

[57] Mondal P, Chakrabarty A. Slotted waveguide antenna with two radiation nulls. IEEE Transactions on Antennas and Propagation. 2008;**56**(9): 3045-3049

[58] Lu X, Wang X, Lu W. A low-profile parallel plate waveguide slot antenna array for dual-polarization application. In: 2017 11th European Conference on Antennas and Propagation (EUCAP). Paris, France: EUCAP; 19-24 Mar 2017. pp. 970-972

[59] Fan G-X, Jin J-M. Scattering from a cylindrically conformal slotted waveguide array antenna. IEEE Transactions on Antennas and Propagation. 1997;**45**(7):1150-1159

[60] Tomar K. Sanjay and Malik, Hitendra, "Density modification by two superposing TE10 modes in a plasma filled rectangular waveguide". Physics of Plasmas

[61] Silvio H, Jankovic G. Scanning Leaky-wave Antenna based on a Waveguide filled with Plasma-like ENG Metamaterial. In: 2006 - MELECON, IEEE Mediterranean Electrotechnical Conference. Malaga, Spain: IEEE; 2006. pp. 280-283

[62] Hirani RR, Pathak SK, Shah SN, Sharma DK. Dispersion characteristics of dielectric tube waveguide loaded with plasma for leaky wave antenna application. AEU – International Journal of Electronics and Communications. 2018;**83**:123-130

[63] Anderson T, Alexeff I. Ân′ Reconfigurable electromagnetic plasma waveguide used as a phase shifter and a horn antenna US6812895B2. 2004

[64] Esteban J, Camacho-Penalosa C, Page JE, Martin-Guerrero TM, Marquez-Segura E. Simulation of negative permittivity and negative permeability by means of evanescent waveguide Modes-theory and ex- periment. IEEE Transactions on Microwave Theory and Techniques. 2005;**53**(4):1506-1514

[65] Barro OA, Himdi M, Lafond O. Reconfigurable patch antenna radiations using plasma faraday shield effect. IEEE Antennas and Wireless Propagation Letters. 2016;**15**:726-729

[66] Barro OA, Himdi M, Lafond O. Reconfigurable radiating antenna array using plasma tubes. IEEE Antennas and Wireless Propagation Letters. 2016;**15**: 1321-1324